AF710145

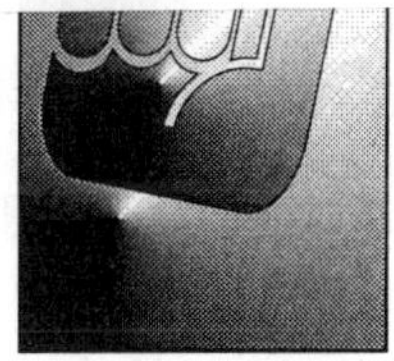

Safety for the Indoor Environment Technician

By

LEO A. MEYER

H. LYNN WRAY, P.E., Technical Advisor

INDOOR ENVIRONMENT TECHNICIANS LIBRARY

LAMA BOOKS

FOREWORD

You are probably working as a technician in one of the indoor environment fields. This means that you have probably heard safety rules many times. However, don't fall into the trap of thinking, "I know all this stuff."

Read each chapter. Then do the Review. In my experience, every time I studied material I "knew all about," I learned new ideas and corrected misunderstandings.

If you study each chapter carefully, you will gain new ideas. More important, you will give yourself a solid understanding of how to protect yourself on the job.

Indoor Environment Technician's Library

This book is part of the *Indoor Environment Technician's Library.* These are practical books that you can use for training or as reference. These books apply to all areas of the indoor environment industry, such as:

Heating, ventilating, and air conditioning
Energy management
Indoor air quality
Sheet metal work
Building commissioning
Testing, adjusting, and balancing

If you are a supervisor training others, you will find that the *Indoor Environment Technician's Library* can make it easier.

Leo A. Meyer

LAMA Books
Leo A. Meyer Associates Inc.
20956 Corsair Blvd.
Hayward CA 94545-1002
888.452.6244
510.785.1099 Fax

ISBN 0-88069-024-0

© 2000 by Leo A. Meyer Associates Inc. All rights reserved. No part of this publication may be reproduced, stored in an electronic retrieval system or transmitted in any form or by any means—electronic, mechanical, or otherwise—without the prior permission of the publisher.

TABLE OF CONTENTS

This manual is intended for safety training and as a reference for general safety rules for work on new construction and on existing buildings. Working safely requires knowledge of proper safety procedures plus judgment and common sense in applying these procedures to the particular job conditions.

LAMA Books specifically disclaims any and all liability for damages of any type whatsoever that may result directly or indirectly from a person's reliance upon or use of the information contained in this manual.

Safety is a broad and complex subject. This manual covers the processes and procedures commonly found in the shop or on the job. It cannot cover all the details of each. Consult the current edition of OSHA regulations in Code of Federal Regulations Title 29, Part 1926, for complete safety regulations for any particular process.

1 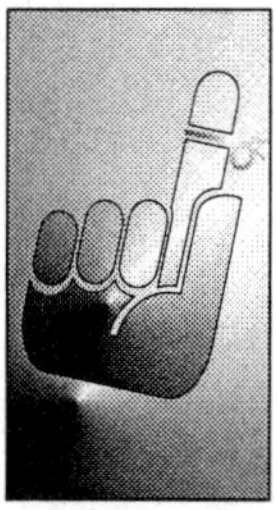 KEEPING SAFETY IN MIND

BASIC SAFETY

Safety is no accident. The most important decision you can make about your job is that you will work safely. This may seem simple. However, you must develop a constant attitude that safety comes first. There can be no occasional lapses. One lapse can mean your death or an injury that can cripple you for life.

This book is based upon the regulations in the book *Code of Federal Regulations 29, Part 1926, Revised July 1, 1997.* This is prepared by **OSHA**, the **Occupational Safety and Health Administration**. Refer to this book for more detail on safety codes. Always use the latest book of regulations available.

Because job conditions vary so much, you must use common sense and good judgment when applying safety regulations to specific situations. However, never violate OSHA regulations. When you use judgment, always choose the safest way to do the job. Most cities have an OSHA office that can be consulted in doubtful situations.

Worker Dies

A worker was fatally injured at a local rolling mill last night. The victim was doing maintenance work on machinery. When power to the machine was turned on, a large metal gate dropped, broke its safety chains, and fell on the victim, crushing his head.

Newspaper clipping

This book is written for technicians working in all branches of the indoor environment, including HVAC, service work, IAQ (indoor air quality), TAB (testing, adjusting, and balancing), and sheet metal shop work.

Working Alone

An indoor environment technician often works alone in a building. Be aware that there is a constant danger of being injured or being trapped in places such as a fan plenum. Always make sure to check in and out with someone in the building so that someone will know if you are missing and will search for you. Carry a cellular phone with you, so you can call for help in emergencies.

Never Take a Chance to Save Time

When you take a chance on the job, you are betting the few minutes you save against a chance of death or a lifetime of physical disability. You often win the gamble, but if you lose only once, you may have to pay for the rest of your life.

Safety is Invisible

Safety is invisible because you never see the accidents that you avoid. You may save many people from death or injury and never know it. For example, when you fasten a piece of plywood over a hole in the roof, you may save someone from a fatal fall. If you pause to pick up a piece of scrap from the shop floor, you may save someone from tripping and breaking a hip. You may say "That doesn't happen often." That's true, but the rare happenings are the ones that kill and maim.

Look Out for Others

Your safety attitude doesn't apply just to you. You must apply it to everyone you associate with. Covering a hole and picking up scrap are ways of looking out for others. When doing such tasks as grinding or welding, you must be aware of the safety of everyone around you. Alert others to any potential hazards.

Keep Your Mind on the Job

Always keep your mind on the job. This is especially important when you are on a repetitious job, or a job that you know so well that it takes little thought. It is easy to let a radio distract you. It is easy to let your mind wander to what you plan to do in the evening or to a personal problem. This is setting yourself up for an accident. Keep your mind on your work—be aware of what your next move is and where your hands are. Always be thinking of the next step of the job and where the hazards are likely to be.

Safety means looking ahead. Plan your work and determine how you can avoid any hazards.

No Horseplay

It is good to enjoy your work and your friends on the job. A laugh makes work go more easily. But jokes must be in words, not in actions. Horseplay or practical jokes have no place on the job. They can lead to serious accidents. Save your clowning around until you are off the job.

DRESS FOR SAFETY

Your work clothing and gear can protect you in many ways. Think about the safety hazards on your particular job and dress accordingly.

Eye Protection

You will hear this repeated many times because it is so important: **Wear eye protection**. In your work in the indoor environment industry, your eyes are often in danger. Oil splashes, sparks from electrical shorts, hot water from a hydronic system, chips of metal from grinding, and sparks from welding are typical eye hazards. Get a good pair of safety glasses that are comfortable and easy to wear. Use these all the time. Change to safety goggles with side shields whenever there is a more serious hazard to your eyes. Sometimes even goggles are not enough. Change to a face shield, which protects the entire face from such things as splashing liquid and sparks from welding or grinding.

Ear Protection

Noise on the job is a long-term hazard that can lead to hearing loss. Constant noise, occasional very loud noise, or vibration can all be damaging. Wear ear protection if sound is a problem. This can be ear-muff type protection or ear plugs. The best ear protection is individually fitted. Ear protection must be designed to reduce noise levels but still allow you to hear higher level warning signals, such as the signal for a truck backing up.

It is easy to ignore hearing protection because the damage is in the future, but it is foolish. You will regret it when you find that you cannot hear changes in the sound of your car engine, or when you cannot take part in ordinary conversation.

Foot Protection

Wearing low shoes with soft tops (such as sport shoes) on the job is foolish. Nothing will protect your feet better than ankle-high leather shoes, with thick non-conducting soles:

- ❑ High tops support your ankles and protect your feet from sparks and sharp objects.

- ❏ Thick soles protect you from nails and other sharp objects.
- ❏ Non-conducting soles protect you from electrical shock.
- ❏ Good soles prevent slipping.
- ❏ Some jobs require steel-toed shoes to prevent feet from getting crushed.

Protective Clothing

Wear work clothes that are **not flammable**. Many modern synthetic materials can burst into flame easily from a single spark, especially if the cloth has absorbed flammable liquids or fumes. Cotton work clothes are safer.

Avoid loose fitting clothes that can catch in machinery or catch on obstructions.

Shark from "Jaws" Claims new victim

HOLLYWOOD (UPI)—The mechanical shark that terrified theater-goers in the movie "Jaws" and now leaps from a lagoon and snaps at tourists visiting Universal Studios sent a workman to the hospital today.

Barry Alkira, 30, a painter at the studio, was working on the fin of the life-like shark when the mock sea creature was "inadvertently activated," a Universal spokesman said.

Alkira was apparently thrown through the air and dragged by the "shark." He was taken by ambulance to St. Joseph's Medical Center in Burbank with possible back injuries.

Newspaper clipping

Long sleeves protect you from welding and grinding sparks, and from hazardous fumes and liquids. They also protect you from long exposure to sunlight, which can lead to skin cancer and other problems. However, long sleeves are a hazard around rotating equipment such as V-belt drives. Wear short sleeves or roll up your sleeves around such equipment.

Other Protection

Hard hats are required on every construction job.

Gloves protect your hands. Select them to fit the job. For example, do not wear common work gloves when handling caustic liquids. They will provide little protection. Instead use gloves made of material that will withstand the chemical. Neoprene gloves are a common type. However, other types may be required for some materials.

Never wear gloves when working around moving machinery. It is too easy for the tips of the glove to be caught in the machinery so that your hands are pulled into the machine.

Remove **rings and other jewelry** when welding, or when working on moving machinery or electricity. A ring can burn your finger badly if it gets hot from electric arc welding. An electrical short can heat a ring and cause a severe burn. Rings, jewelry, loose clothes, ties, and long hair pose a serious danger of getting caught in machinery and pulling you into it. You can still wear long hair—just keep it tucked up under a cap or by other means.

HOUSEKEEPING

One of the best ways to improve safety is to **keep the work area clean**. Studies have shown that messy work areas have a higher rate of accidents than clean ones. Such things as metal scraps, packing materials, and general debris from work must be picked up immediately. There is usually a trash can nearby. If there is not, use a cardboard box, or put debris in an out-of-the way corner until it can be picked up.

Clean up oil spills or other spills at once to prevent falls.

Don't leave tools and equipment lying around. They create a tripping hazard. Never leave tools or materials on ladders or scaffolds. If a ladder is moved, the tool will fall off and could hit someone. Loose tools on scaffolds can be kicked off. In addition, loose tools can result in loss and can waste time in trying to find the right tool.

EMERGENCY PROCEDURES

Be aware of these facts: Rescuers are often killed or injured. Untrained rescuers often cause further injuries to a victim.

When a person is injured on the job, every situation is different and you have to use your best judgment in deciding how to make the rescue. However, you should follow these general guidelines:

- ❑ Call for professional medical help immediately. On a construction site, emergency numbers are posted. However, an indoor environment technician often works in an occupied building. Have emergency numbers written on something that is always with you, such as your tool kit or clipboard.
- ❑ Keep the victim warm and lying quietly.
- ❑ Have someone stay with the victim to provide reassurance and comfort.
- ❑ Find someone on the job who is trained in first aid. Unless you are trained, provide only the first aid that is absolutely necessary, such as stopping severe bleeding.
- ❑ If you must rescue a victim from an ongoing danger (such as electrocution or drowning), think before you act. Do not become a victim yourself.

SAFETY AS A TEAM EFFORT

Working safely is a team effort. The main members of the team are you, your employer, employee organizations such as unions, and government safety organizations. Consider yourself the lead member of the team. Unless you have an attitude of safety and follow all safety rules, the work of everyone else on the safety team is wasted. You have the first responsibility for safety for yourself and your fellow workers.

YOUR RIGHTS

The U.S. government, through OSHA (Occupational Safety and Health Administration), says that all workers are entitled to safe working conditions. By law, you can refuse to work where there is a serious hazard. However, you must be willing to do other work or do the original job when conditions are made safe.

Achieve safety on the job without damaging working relations with your fellow workers, supervisors, and employers. Most dangerous situations develop gradually. If you see a possible danger, point out the problem to your supervisor and find out how the job can be made safe.

If there is a continuing problem, you may have to take further action. If you work in a union shop, have the shop steward or a union official work out the problem with supervisors and employers. If necessary, you can appeal to the local OSHA office.

Try not to let a situation develop to the point that the only way out is a flat refusal to do the work. However, if you have a work situation that means a real risk of injury or death, it is only common sense to refuse to do the job until it is safe.

YOUR RESPONSIBILITIES

Safety rules on the job are great, but they don't mean a thing unless you follow them and refuse to work under unsafe conditions. You must take responsibility for your own safety and for your fellow workers:

- ❑ **Do not tolerate drugs or alcohol in others on the job.**
- ❑ **Never be under the influence of drugs or alcohol on the job.**

- ❑ Look out for the safety of everyone on the job.
- ❑ Don't try to do a job by yourself if it takes more than one person to do it safely.
- ❑ Report unsafe working conditions.
- ❑ Report defective tools and equipment.
- ❑ Report all injuries, no matter how small. A small injury can develop into a serious one. You want a record that the injury occurred on the job.

YOUR EMPLOYER'S RESPONSIBILITIES

Most employers want safe working conditions simply because they have an interest in their employees and want them to be safe. An employer also has solid financial reasons for keeping safe working conditions. An accident or death on the job can result in a substantial fine from OSHA. In addition, every reported accident increases insurance rates. Every employer must carry Workers' Compensation Insurance, which provides benefits for on-the-job accident victims. The insurance rates depend upon the safety record of the company. Every accident results in an increase in rates. A bad safety record can increase the insurance rates so much that a company has trouble competing for work.

OSHA spells out the employers' obligations for safety. The employer must:

- ❑ Maintain a safe working environment.
- ❑ Develop a safety training program.
- ❑ Have emergency phone numbers posted on the job.
- ❑ Have a first aid kit available on the job site.
- ❑ Have at least one person on the job who is trained in first aid.
- ❑ Enforce safety rules on the job site.

- ❏ Fix unsafe conditions.
- ❏ Investigate accidents.

If your employer is a subcontractor on a project, these safety requirements are usually shared with the general contractor or with other subcontractors. For example, not every subcontractor must have a person trained in first aid on a construction site. (Usually the general contractor has such a trained employee on the job.) However, **every** employer must maintain safe working conditions and have a safety training program.

OSHA

The main enforcer of safety on the job is **OSHA** (**Occupational Safety and Health Administration**). Some states have an agency that is the equivalent of OSHA. State safety rules must be at least as strict as federal OSHA, and they may be stricter.

There is usually a state or federal OSHA office near you. Local OSHA offices make on-the-job inspections for safety conditions. If OSHA finds there are safety violations, the employer will be given notice to correct the conditions and may be fined. However, most government safety organizations have a very heavy work load, so they cannot inspect every workplace frequently.

It comes back to the fact that **you** must take the major responsibility for your safety.

> A worker was repairing an air-operated valve that he had turned off but not disconnected and locked out. During the repair he slipped and accidentally turned on the switch, which let air into the valve. His hand was caught and crushed in the valve.
>
> **This would not have happened if he had disconnected the power source and locked it out.**

2 HAZARDOUS MATERIALS

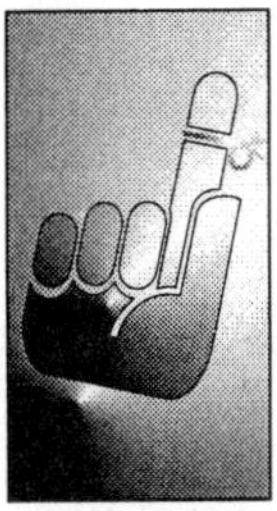

Hazardous materials can injure your health. They may be solids, liquids, fumes, explosives, and even water.

Hazardous solids include lead and asbestos. Particles that enter your body can affect your health many years later. Old insulation on piping, boilers, and furnaces may contain asbestos.

Hazardous liquids include solvents, gasoline, cements, corrosives, and glues. All of these are harmful if taken internally. Many of them can also harm your skin. Some, like gasoline, are very flammable. Many are also harmful because they give off hazardous vapors. Mercury (from thermometers) is also hazardous. Avoid contact with it.

Hazardous vapors are produced by gasoline engines, boiler and furnace vents, welding, and other processes. Some CFC refrigerants can create a toxic gas in the presence of a flame. Some newer refrigerants in centrifugal chillers are hazardous.

Explosives are hazardous of course. LPG (liquefied petroleum gas) such as propane and butane can be especially dangerous because the fumes are heavier than air and they stay near the floor where a spark can ignite them. Concentrations of dust from work areas, grain, and other materials are also very explosive.

Water that is standing or constantly recirculated may be hazardous because it can breed organisms that cause diseases, such as Legionnaires' disease. Water in condensate drain pans, cooling towers, and open piping systems may be hazardous.

Figure 1 lists some of the common hazardous materials found on job sites.

SHORT TERM AND LONG TERM HAZARDS

Short term hazards cause harm within minutes, hours, or days. **Long term hazards** may not show any health effects for years. They have one thing in common—they harm you or kill you. The only difference is the time it takes to do it.

Examples of short term hazards are vapors and fumes from sprays, acids, and gasoline engine exhausts in a confined space. Such things can affect your breathing, or make you sick, or kill you.

This list of common hazardous materials **is not complete**. You will encounter other hazardous materials on the job.

- Paint
- Lead
- Engine exhausts
- Gasoline
- Acids
- Solvents
- Asbestos-containing materials
- LPG (liquefied petroleum gas)
- Paint removers
- Dust
- Hot asphalt
- Gas appliance vents
- Welding gases
- Liquid cleaners
- Glues

Fig. 1: Hazardous materials commonly found on the job

Examples of long term hazards are exposure to asbestos and lead. If you handle these materials, you will not feel any immediate effects. However, lengthy exposure to lead can affect you months or years later. Exposure to asbestos can cause serious disease many years later.

PROTECT YOURSELF

You are the most important person to you and your family. That's why **you** have to take responsibility for your own safety on the job. You probably are the only one who knows the changing job conditions. An indoor environment technician often works on the job without a foreman or supervisor. You must make sure that working conditions are safe for you. For example, if you must work in a confined space, it is a temptation to say it is too much trouble to get ventilating equipment and someone to stand by—and anyway, you will only be in there for a short time. Many workers have died because of this kind of thinking.

You must also protect your family. For example, lead or asbestos on contaminated work clothes can expose family members. Children especially may be harmed by these materials. Contaminated work clothes should be washed separately.

EMPLOYER RESPONSIBILITIES

Employers must follow the many safety codes set by OSHA (Occupational Safety and Health Administration) regarding hazardous materials.

Employers must have a safety training program for employees. They must inform workers when hazardous materials are being used and have an MSDS (material safety data sheet) on file for every hazardous material used. If necessary, employers must provide individual respirators for employees and train them how to use them.

MATERIAL SAFETY DATA SHEETS (MSDS)

Every hazardous material is supposed to have all safety information on the container. Check the manufacturer's information on the container and on any pamphlets provided. If this does not provide all the information you need, then consult the MSDS.

An **MSDS** is a **material safety data sheet**. These are data sheets provided to the customer by the manufacturers of materials used in industry and business. Figure 2 shows part of a typical MSDS. The format of these sheets varies, but they all provide the same kind of information.

Some of the information is highly technical and is used by specialists such as industrial hygienists. However, much of the information is useful for anyone who uses the material. Look for information like the following:

- ❑ How the material is hazardous
- ❑ How to treat exposures to the materials
- ❑ How to clean up spills
- ❑ Whether the material is flammable or explosive
- ❑ Type of respiratory protection required

OSHA requires that employers provide information about hazardous materials by means of a hazard communication program, labels, MSDS, and information and training. An employer is expected to keep an MSDS on every hazardous material used.

If there is any doubt about a material, you have the right to request an MSDS from your employer. However, as with every safety regulation, use your rights with common sense. NEVER work under hazardous conditions. But do not abuse your rights by unreasonable requests.

Reading an MSDS

The highly technical information on an MSDS may seem confusing, but there is important information that you can use. The wording may vary, but all sheets contain the following information:

- ❑ The **date** of the MSDS (Fig. 2)—If the date is before 1983, the MSDS is out of date and may not contain everything that is presently required.

MATERIAL SAFETY DATA SHEET

I. IDENTIFICATION

Manufacturer's name: IET Chemical Company
Address: Centerville, USA
Product: Acetone X
Emergency phone: 800-555-8888
Date prepared: 01/01/99

Chemical name: Acetone
Generic name or trade name: Dimethyl, Ketone, Propanone
Other identifying information: CH_3COCH_3
Hazard severity: Extremely flammable

II. HAZARDOUS INGREDIENTS

Material: Acetone **Percentage of mixture:** 99.5%
Nature of hazard and toxicity data: Toxic, respiratory irritant

III. PHYSICAL DATA

Boiling point (°F): 133°
Vapor density (air = 1): 2.0
Melting point (°F): N/A
% volatile by volume: 99.5
Vapor pressure (mm Hg): 25°C - 226
Water solubility: Complete
Specific gravity: 0.7917
Evaporation rate (nBuAc=1): 5.6

Appearance and odor: Clear, colorless, volatile liquid with a fragrant mint-like odor. Unfatigued odor recognition threshold (100% of test subjects) is 100-150 ppm. Distinct odor at 680 ppm.

IV. FIRE AND EXPLOSION HAZARD DATA

Flash point (°F): 0°F **LEL:** >2.1 % **UEL:** 12.8%
Method used: Closed cup
Extinguishing agents: Alcohol foam CO_2, dry chemical
Fire or explosive hazards: Dangerous fire hazard and moderate explosion hazard when exposed to heat and flame.
Special fire fighting equipment: Use self-contained breathing apparatus.
Special fire fighting procedures: Class B fire. Use a blanketing effect to smother flame.
Other: Water spray may not be effective in extinguishing. Use water to cool fire-exposed containers and to dilute and reduce fire intensity.

V. HEALTH HAZARDS

May cause eye injury. May be harmful if swallowed. May be harmful if

Fig 2: Part of a typical MSDS

II. HAZARDOUS INGREDIENTS

Material: Acetone **Percentage of mixture**: 99.5%
Nature of hazard and toxicity data: Toxic, respiratory irritant

Fig. 3

- **Hazardous ingredients** (Fig. 3)—This section may include hazard information as well as ingredients.

III. PHYSICAL DATA

Boiling point (°F): 133°
Vapor density (air = 1): 2.0
Melting point (°F): N/A
% volatile by volume: 99.5
Vapor pressure (mm Hg): 25°C - 226
Water solubility: Complete
Specific gravity: 0.7917
Evaporation rate (nBuAc=1): 5.6

Appearance and odor: Clear, colorless, volatile liquid with a fragrant mint-like odor. Unfatigued odor recognition threshold (100% of test subjects) is 100-150 ppm. Distinct odor at 680 ppm.

Fig. 4

- **Physical data** (Fig. 4)—This includes the evaporation rate of the material. An evaporation rate over 3 means that the material may produce fumes to a hazardous level quickly.
- **Fire and explosion hazard data** (Fig. 5)—This gives the flash point, which is the temperature at which the material may ignite. A **low flash point** means that the material is **highly flammable**. (Gasoline has a flash point of -50°F.) This section also describes appropriate fire fighting methods.

IV. FIRE AND EXPLOSION HAZARD DATA

Flash point (°F): 0°F **LEL**: >2.1 % **UEL**: 12.8%
Method used: Closed cup
Extinguishing agents: Alcohol foam CO_2, dry chemical
Fire or explosive hazards: Dangerous fire hazard and moderate explosion hazard when exposed to heat and flame.
Special fire fighting equipment: Use self-contained breathing apparatus.
Special fire fighting procedures: Class B fire. Use a blanketing effect to smother flame.
Other: Water spray may not be effective in extinguishing. Use water to cool fire-exposed containers and to dilute and reduce fire intensity.

Fig. 5

V. HEALTH HAZARDS

May cause eye injury. May be harmful if swallowed. May be harmful if inhaled. May cause skin reaction. May be harmful if absorbed through skin.

TLV: 750 ppm
PEL: 1000 ppm
Routes of exposure: Oral, skin contact, inhalation
Acute symptoms: Eye irritation and damage. Mucous membrane irritation.
Chronic symptoms: Dermatitis, headache, nausea, dizziness, narcosis
First aid: Eye contact—flush promptly with plenty of running water for 15 minutes and contact a physician. Skin contact—wash off with water, remove contaminated clothing, get medical assistance if irritation persists. Inhalation—remove to fresh air; contact a physician if effects are more severe than a headache; restore or support breathing if required. Ingestion—give water to dilute. Induce vomiting and contact a physician.

Fig. 6

- **Health hazards** (Fig. 6)—This tells you what first aid measures to take. The term **PEL** means **permissible exposure limits**. This means you are not allowed to be exposed to more than specified PEL amount.

VII. SPECIAL PROTECTION INFORMATION

Respirator: Half-mask cartridge **Chemical splash**: Goggles, gloves, apron

Ventilation: Local exhaust, mechanical general exhaust, explosion-proof

Fig. 7

- ❑ **Special protection information** (Fig. 7)—Tells you what personal protective equipment (gloves, face shield, respirator, etc.) is required.

VIII. SPILL OR LEAK PROCEDURES

Specific steps: Provide maximum explosion-proof ventilation and eliminate all sources of ignition. Flush with water spray to clear area of acetone. (Flush to open ground, not to drains, sewers, or surface water.) Or dilute with water and pick up with non-sparking tools for disposal.

Disposal methods: Incineration is recommended. Follow federal, state, and local regulations. Consider reclaiming by distillation.

Other precautions: Protect against contact with liquid and inhalation of vapors during clean-up.

Fig. 8

- ❑ **Spill or leak procedures** (Fig. 8)—This is information that may be needed in case of an accident.

CONFINED SPACES

Hazardous materials are an extreme hazard in **confined spaces** that have no outside air circulation. These are such areas as tanks, service tunnels, pits, trenches, boilers, ducts, and unventilated attic crawl spaces. The danger in confined spaces may be due to:

- ❑ Hazardous vapors or fumes—Breathing them is harmful.
- ❑ Lack of oxygen—If a gas that is usually harmless builds up so that it replaces the oxygen in the air, it can cause a person to collapse and even suffocate.
- ❑ Explosive gases—Even a small spark can cause them to explode.

Be cautious about any area where vapors and fumes are present. For example, gas engines that are running are dangerous in areas even partly confined. They produce carbon monoxide (CO), which can kill, and they use up the oxygen in the air. Welding often produces dangerous fumes. Chemical processes should be checked for dangerous fumes. These are only a few common examples. You must constantly evaluate areas to decide whether there is danger.

A **safety monitor** is another worker stationed outside the confined space. The safety monitor tends a safety rope and watches to make sure that the worker in the confined space is safe. There should be other workers in the area because it may require more than one person to pull a victim out of danger. The safety monitor should not enter the confined space for a rescue without taking all the required safety precautions. This includes a lifeline and another safety monitor.

A common tragedy is that a worker enters a tank to clean or repair it without a safety monitor. The worker may become unconscious before becoming aware of what is happening and never calls for help. Another common tragedy is that a safety monitor enters the space to make a rescue, but does not notify anyone before entering the space. Both the worker and the monitor are overcome.

Always treat enclosed tanks as a confined space. Many workers have suffocated in empty tanks. Even if they are empty, dangerous vapors from the former contents may remain. Even if there is no danger of vapors, you may be overcome by the lack of oxygen in the air.

The employer is responsible to see that a worker entering a confined space is instructed about the hazards and about the safety precautions to be taken:

- ❑ Be sure of a **supply of air**. You may need a respirator. Or you may only require a fan that delivers outside air for ventilation. The fan must be connected to a length of flexible duct so that it is delivering clean air, not just recirculating the polluted air from the confined space.
- ❑ A **safety monitor** should be stationed outside the confined space.

PREVENTIVE GEAR

Chapter 1 described the protective clothing such as long sleeves, high-top shoes, gloves, and eye protection. These are all important when working with hazardous materials. Special eye protection and gloves are usually required for hazardous materials. For liquids, **splash-proof goggles** may be required. They have side shields to prevent drops of hazardous liquid from entering the eyes. Gloves have to be of the proper materials. For acids, usually a neoprene glove is required. Consult the manufacturer's data for the proper type of glove to be used.

Respirators are required when working with many hazardous materials. Respirators keep harmful material out of the respiratory system. Respirators are not always comfortable to use, but it is very important to use this protection and to use it properly in order to protect your health.

A respirator has to be **fitted** to the person and then tested to be sure it fits properly so that there are no leaks. The employer must provide properly fitted respirators, to provide training in their use, and to maintain them. The following regulations apply:

- ❑ The worker must have a medical evaluation before using a respirator.
- ❑ The worker who uses a respirator must not have facial hair.
- ❑ The employer must provide annual training in the use of the respirator and an annual fit test.

After the respirator is individually fitted and tested, only that one person should use it. Respirators should be cleaned and disinfected after every use.

There are different types of respirators. The wrong type for the particular work situation may not protect you. The two basic types of respirators are **air-purifying** and **supplied-air**.

Air-Purifying Respirators

Air-purifying respirators (Fig. 9) purify the air by filtering, adsorbing, or neutralizing harmful ingredients in the air before they enter the respiratory system. It is very important to choose the correct filter or cartridge for the nature and amount of the hazard in the air:

COURTESY LAB SAFETY SUPPLY

Fig. 9: Air-purifying respirator

- ❑ Mechanical filters are used with these respirators to trap particles in the air. These filters need to be changed when breathing becomes more difficult because the filters are becoming clogged.

- ❑ Chemical cartridges are used to remove gases and vapors. They work by chemically neutralizing the fumes or by adsorption (attracting and holding them on the surface of a substance). They need to be changed when the wearer notices odors, tastes, or irritation.

Color coding may be used to distinguish different types of air-purifying cartridges, such as:

- ❑ Black – Organic vapors (gasoline and many solvents)
- ❑ White – Acid gases (specific to chlorine and hydrogen chloride)
- ❑ Yellow – Combination of organic vapor and acid gas
- ❑ Green – Special applications (ammonia, iodine, pesticides, formaldehyde, etc.)
- ❑ Orange – Dust, mist, and fumes
- ❑ Magenta (HEPA filter) – Radionuclides, dust, mists, fumes, asbestos

Particulate filters (that trap particles in the air) have the following categories:

- ❑ N – Only approved for areas free of oil
- ❑ R – May be used in an oil aerosol environment but may have use-time restrictions
- ❑ P – Permitted in any environment with no time restrictions

Each of these three categories has three classes based on filter efficiency:

- ❑ 95%
- ❑ 99%
- ❑ 100% (actually 99.97%)

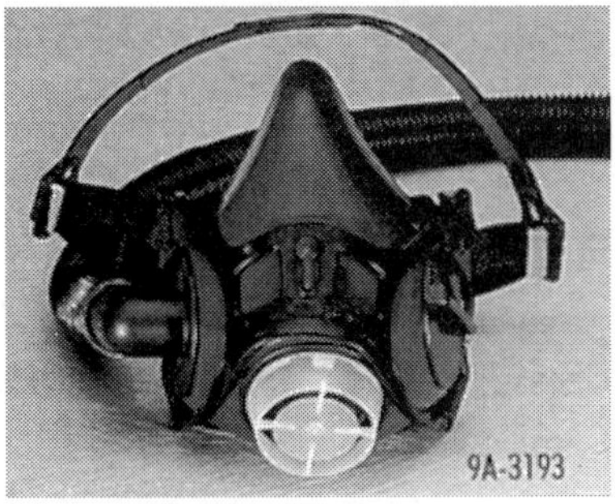

COURTESY LAB SAFETY SUPPLY

Fig. 10: Supplied-air respirator

Supplied-Air Respirators

Supplied-air respirators (Fig. 10) supply the air you breathe in order to block out the surrounding air completely. They are used if air purifying respirators will not be effective in the situation. The user breathes clean air supplied from an outside source:

- ❏ The supplied air may be delivered by a hose from an outside source. This may be called an **air-line respirator**.
- ❏ The supplied air may be carried in tanks on the user's back. This may be called **self-contained breathing apparatus** (**SCBA**).

There are many variations of these types of respirators, and suppliers have different terms for them.

NOISE

Noise is a long-term hazard, even though it is not a material. Our ears cannot survive the clanking, screeching, thumping, and pounding of modern industry. Continual exposure to noise will lead to hearing loss. It may leave you deaf to higher level sounds so that listening to conversation becomes difficult. It can also cause a constant ringing in your ears that can be very unpleasant.

Hearing loss can be caused by very loud noises. But even sound that is not much higher than normal can damage your hearing if you are exposed to it for a long time.

Sound level is measured in **decibels** (**dB**). Sound that is over 90 dB can damage your hearing over a period of time. At 130 dB, sound is painful and is certain to damage hearing. Figure 11 shows the decibel level of various sounds.

140 dB	Jet plane, gunshot
130 dB	Riveting steel tank
120 dB	Auto horn
110 dB	Power saw, rock band
100 dB	Punch press
90 dB	Heavy truck
80 dB	Alarm clock, restaurant
70 dB	Conversation
60 dB	
50 dB	
40 dB	
30 dB	Soft whisper

Fig. 11: Approximate sound level in decibels (dB)

OSHA has limits for the time that workers can be exposed to sound that is over 85 dB. The employer is responsible to try to reduce noise levels if this is practical. Sometimes this can be done by setting up barriers, by muffling the noisy equipment, by using noise absorbing materials, or by other means. When noise cannot be reduced to a safe level, the employer must provide ear muffs or ear plugs:

- ❑ **Ear plugs** are made of a soft material that will conform to the shape of the wearer's ear canal. Ear plugs are discarded after each use.
- ❑ **Ear muffs** must have a tight seal with the head. Long hair, glasses, and other obstructions may make them less effective. They are used with high level noises.

When there is danger of damage to your hearing from noise, be sure to use the ear protection available to you.

REVIEW

1. Name two hazardous solids.
2. Name four hazardous liquids.
3. Name two hazardous fumes.
4. What government agency sets codes regarding hazardous materials?
5. What do the letters MSDS stand for?
6. List three dangers of working in a confined space.
7. What is the term for a worker who is stationed outside a confined space to rescue the worker inside if necessary?
8. What type of respirator uses filters or cartridges to purify the air the worker breathes?
9. What type of respirator provides the worker with a supply of air from an outside source?
10. Name two types of ear protection you can wear.

Check your answers in the back of the book.

3

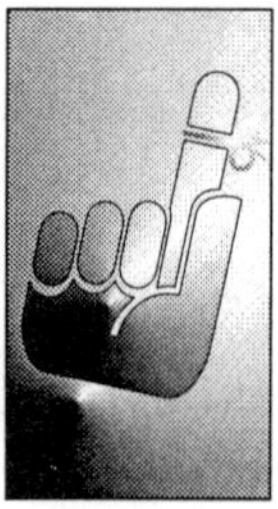

ELECTRICAL SAFETY

THE DANGERS OF ELECTRICITY

You use electricity every day of your life, so it is easy to take it for granted. Yet electricity is a major cause of injury on the job. An average of 1200 people are killed every year by electricity. Many more suffer painful shock.

A jolt of electricity can hurt you, burn you, knock you unconscious, or kill you. How much electricity will kill you? It's hard to say. An amount that might just give you a slight shock one time might kill you another time. Damp skin, wet ground, metal-tipped shoes, or a two-handed grip can make even a relatively small amount of electricity fatal.

The main danger of electricity is that the voltage clamps your heart muscle and causes a very irregular movement. As a result, your heart stops pumping, blood circulation stops, and you stop breathing. If CPR is started immediately, it may be able to revive someone who has stopped breathing because of an electric shock.

Electric shock can also cause other muscle spasms. If someone touches a live electrical line, a muscle spasm of the hands may make it impossible to let go of the wire. If you are on a ladder or a scaffold, a shock can cause you to fall.

Worker Hit by Electrical Wire

A worker was treated for electrical shock at the hospital after an on-the-job accident. According to the deputy sheriff, the worker was installing drywall when he backed up and an electrical wire hanging from the ceiling hit him on the side of the head.

Newspaper clipping

High voltage can cause burns. Electricity can start a fire. Electrical fires often start out of sight and can become dangerous before they are even noticed.

Electricity is safe only if you use all the safeguards built into the system, as well as common-sense safe working practices. Can you answer these questions?

- ❑ You turn off the power in order to work on a circuit. Yet you get a shock. Why?
- ❑ You turn off the power and remove the electrical connections to a motor before working on the motor. Yet you still get a shock. Why?
- ❑ Someone working on a 110 volt house circuit gets a shock and dies. Why?
- ❑ Your buddy on the job receives a 440 volt shock, is unconscious, and is still connected to the wires. What should you do?

This chapter will give you answers to these questions. For even more information, study the book *Basics of Electricity,* in this series, the *Indoor Environment Technician's Library.* The more you know about electricity, the safer you can be.

ELECTRICAL CIRCUITS

To work safely with electricity, you should know a few things about it. Think of electricity flowing in a wire circuit like water flowing in a pipe. Electricity is safe as long as it is contained in the circuit designed for it.

In a proper circuit, electricity flows from the power source through a conductor to the energy-using device (motor, lamp, heater) and back to the power source (Fig. 1). In a single-phase circuit, one wire connected to the switch for

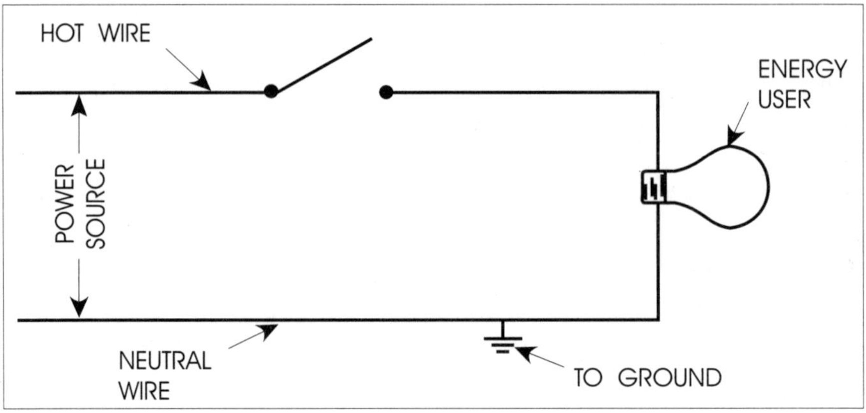

Fig. 1: An electrical circuit has a hot wire and a neutral wire

the energy-user is the hot wire (Fig. 1). The other wire is the neutral wire (Fig. 1). The neutral wire is always **grounded**. That is, it is attached to something that would conduct electricity to the ground where it would be neutralized. The neutral wire is sometimes mistakenly called the ground wire because it is grounded.

Some materials are good **conductors**. That is, they allow electricity to flow through them easily. Silver, copper, aluminum, most other metals, and water are good conductors.

Some materials are good **insulators**. They block the flow of electricity and do not allow it to flow through them easily. Rubber, glass, most plastics, and dry air are good insulators.

Other materials are in between—neither good conductors nor good insulators.

The difference between conductors and insulators makes it possible to use electricity safely by keeping it contained within a circuit. Copper is commonly used for wiring because it is a good conductor at a reasonable price. The wire is protected with insulation that keeps the electricity from escaping to anything else—such as people.

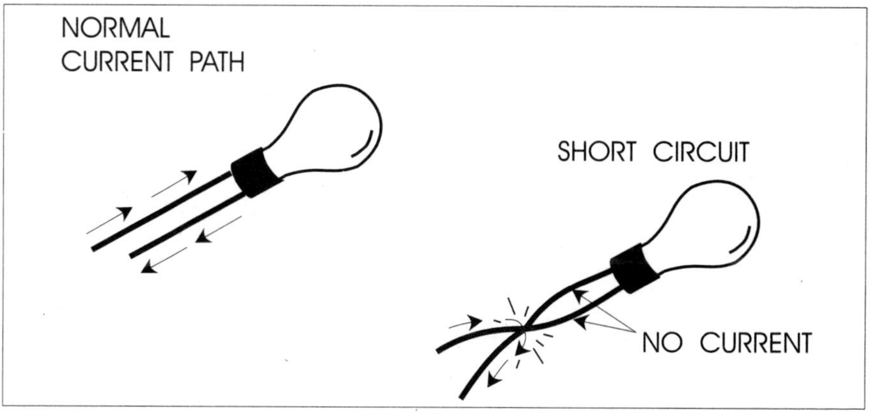

Fig. 2: A short circuit causes high current flow in the conductor

Short Circuits

Trouble starts when electricity escapes from the circuit designed for it. It might happen because the insulation is damaged, perhaps because excessive electric current causes high heat in the conductor and burns through the insulation. It might happen because damaged wires allow the electricity to bypass the energy-using device such as the energy user in Fig. 2. Since the bypass has less resistance than the path through the energy user, more current surges through the short circuit. The high current heats up the conductor and can burn through the insulation. This type of short circuit can cause a fire, explosion, or other damage to people or equipment.

Fig. 3: You are a good path for electricity

Escaped electricity will find a path that leads to the ground where it is neutralized so that it is no longer dangerous. If you become part of that path, the electricity flows through you and into the ground (Fig. 3). You get a shock.

Electricity always takes the path of least resistance. It flows through the best conductor. Unfortunately, the human body is a good conductor, since it is mostly water. If a human body is in contact with the ground, electricity can easily flow through the body to the ground. If it flows through your heart on its path to the ground, your heart may stop (Fig. 4). When you work with electricity, your basic safety practice is to keep from being a conductor to ground.

Fig. 4: If electricity flows through your heart, it may stop

On DC circuits, you must touch both wires to be shocked. However, automotive DC circuits are grounded by being connected to the auto frame and engine. Touching the hot wire and the car frame is the same as touching both wires.

STORED ELECTRICITY

Electricity isn't always moving through wires. It can also be stored in various ways. A battery is a familiar way of storing electricity in a DC circuit. In AC motors and other electrical equipment, a capacitor is a small component that stores electricity. Even when the equipment is completely disconnected from any power source, a capacitor may hold enough electricity to give you a shock if you touch it. You have to know how to discharge the stored electricity in the capacitor by letting it escape to ground. Have a qualified person show you how to do it.

COMMON VOLTAGES

In general, the higher the voltage in a given circuit, the higher the current, and therefore the more dangerous the shock. Human skin usually resists anything under 48 volts, so you probably won't feel a voltage under this amount.

Most power used in homes and small businesses is single phase, 110 to 120 V (volts). However, almost every house has a 220 volt circuit for large appliances such as the washer and dryer. On the job, you may encounter 208V, 220V, 230V, 440V, or 460V. Anything over 600V is considered high voltage and requires special protection devices.

WALL PLUGS

The wall outlets and plugs for 110V or 120V can be one of four kinds:

- ❑ Old style outlet for 2-prong plug—It has no safety feature and it is obsolete. It is seldom seen except on inexpensive extension cords and old outlets.
- ❑ **Polarized 2-prong plug** (Fig. 5)—One flat prong is wider than the other. One slot on the outlet is wider than the other. The plug can be inserted in only one way so that the neutral wire of the circuit is connected to the neutral wire of the equipment and the hot wire of the circuit to the hot wire of the equipment. There is less chance of a short and fewer electrical problems for the equipment. The polarized plug provides some protection from electrical shock, but not as much as the grounded 3-prong plug.

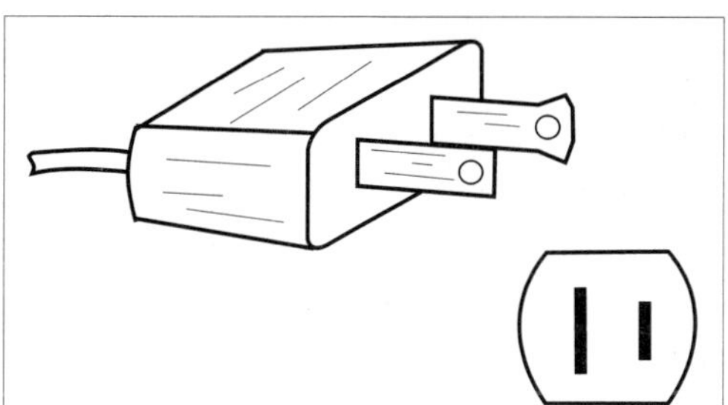

Fig. 5: A polarized plug has one prong wider than the other

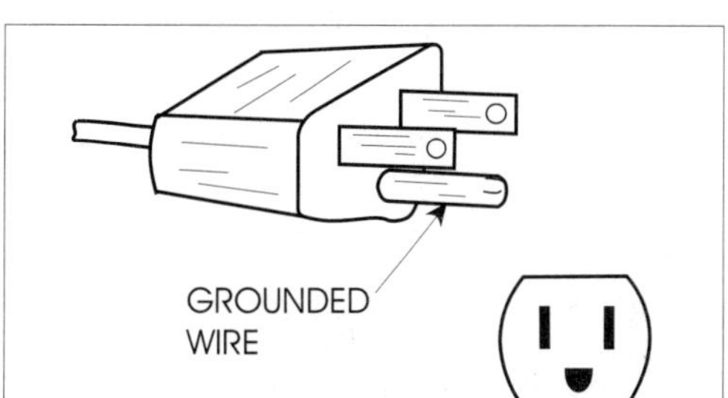

Fig. 6: A grounded plug has one round prong

- ❑ **Grounded 3-prong plug** (Fig. 6)—A round prong is grounded. It is also attached to the frame of the equipment. If there is a short, current will pass through the grounded plug—not through the frame to you. The ground wire that the round plug contacts is separate from the grounded neutral wire of the circuit. A 3-prong plug and outlet is commonly used for 120 volt circuits.

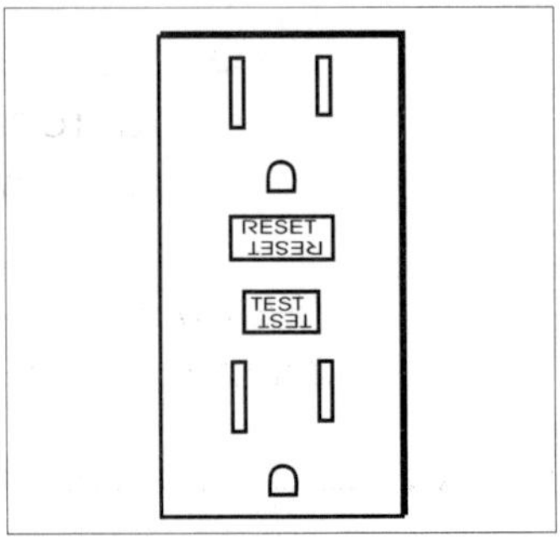

Fig. 7: A GFCI outlet is required for wet conditions

- **Ground-fault circuit interrupter** (**GFCI** or **GFI**) (Fig. 7)—Electrical outlets exposed to the weather or to other wet conditions (such as kitchens and bathrooms) must be GFCI outlets. The GFCI monitors the current in the two wires of the circuit. These two currents should be equal (within a certain tolerance). If the two are not equal, it means there is a short in the circuit, so the GFCI disconnects the circuit. A GFCI outlet has a reset button on it. OSHA requires GFCI outlets on construction sites. The outlet on a GFCI is the grounded 3-prong type.

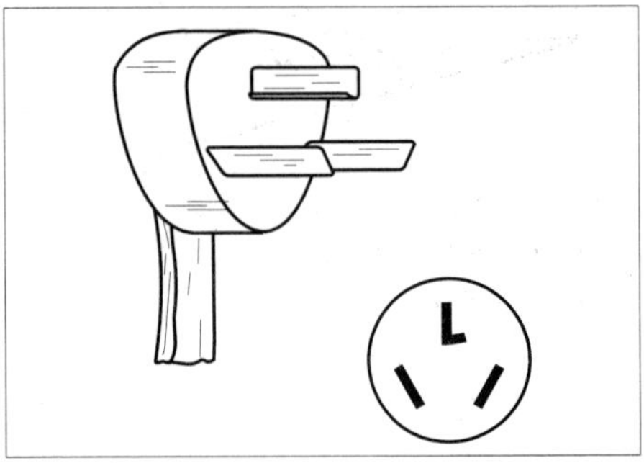

Fig. 8: Plug for 220V single-phase circuit

For 220 volt, single phase circuits, the wall outlet has a special design (Fig. 8) to accept a plug with three flat prongs (or some other special pattern) so that it cannot be confused with a 110 volt circuit.

PROTECT YOURSELF

You must protect yourself from the four main dangers of electricity:

- Shock (ranging from a mild jolt to death)

- ❑ Fire
- ❑ Burns (which may spread beyond the surface burn that shows)
- ❑ Injury (from machinery started up while you work on it; from a shock that causes you to fall into moving equipment or knocks you off a ladder or scaffold)

Fortunately, by following safety rules, you can avoid most electrical accidents. Your first line of defense against electricity is to keep yourself from being a conductor:

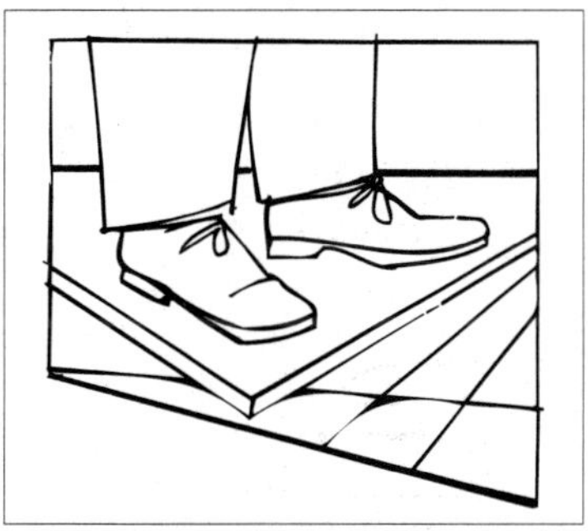

Fig. 9: Stand on a dry board or rubber mat. Wear rubber-soled shoes

- ❑ Keep dry. Avoid working on electrical equipment if you are wet. Never work on an electrical system if your shoes are wet.
- ❑ Don't stand in puddles or on wet ground. If possible, stand on a rubber mat or dry wood platform when you work on electrical equipment (Fig. 9).
- ❑ Wear rubber-soled shoes (Fig. 9).
- ❑ Don't wear metal, such as a watch, ring, or other jewelry. These can cause accidents in three ways: They can catch in machinery; they can conduct electricity; they can cause a severe burn as electricity arcs between the body and the metal.
- ❑ Avoid contact with metal pipes because they provide a perfect ground.
- ❑ Use a wood or fiberglass ladder, not a metal ladder.
- ❑ If you must work around energized circuits, use one hand only (Fig. 10).

Fig. 10: Use one hand when working on live circuits

If both hands are in contact, the electrical path can go from one arm across the heart to the other arm (Fig. 4). If only one hand is in contact, the shock goes from the arm down a leg to the ground (Fig. 3).

KEEP IT TURNED OFF

Never assume that a circuit is turned off. Of course you will turn off the power to any equipment you have to service, but you must make sure it really is off and that it stays off. Too many accidents have happened because one worker turned a switch on after another worker had turned it off for good reason. There are two dangers:

- ❏ Electricity can cause a shock.
- ❏ Equipment can start when you aren't expecting it and can injure or kill.

If you want it off, make sure it stays off.

For equipment on an electric cord, unplug it before you work on it, and put a warning tag on the plug telling others that you are working on the equipment.

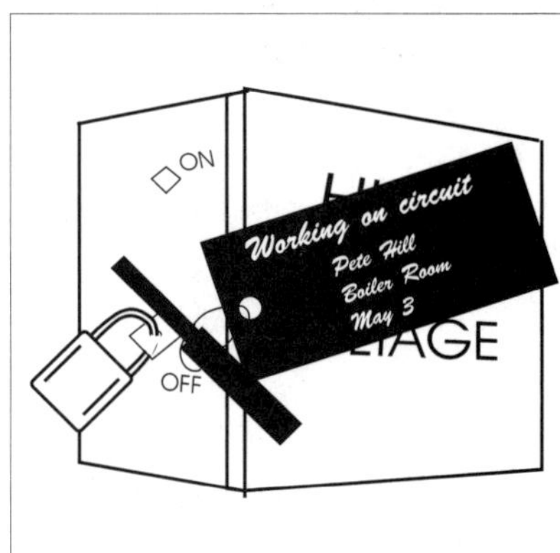

Fig. 11: Lock it out and tag it

For equipment with permanent electrical connections, turn off the power source. Lock it out with your own padlock. Even if a switch is already locked out, put your own padlock on it. Tag the lock with your name, location, and date (Fig. 11).

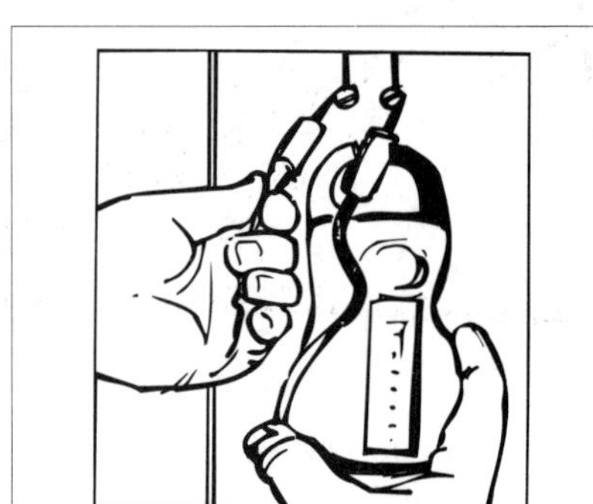

Fig. 12: Take a voltage reading to make sure the circuit is dead

Even if you have turned off the main power or disconnect, take a voltage reading on the lines to be sure they are dead (Fig. 12). If the switch is not properly wired, there could be power in the circuit even though the switch is in the **off** position.

Make sure that any capacitor is discharged safely before you begin work. If you do not know how to do this, get someone to show you.

Remove locks and tags as soon as you are done working. If they are left on longer than needed, other workers won't take them seriously in the future. Make sure you have your own lock and tags.

KEEP EQUIPMENT SAFE

Don't use faulty equipment. If anything electrical—from a small hand tool to a major piece of equipment—is damaged in some way it is not safe. Get it repaired or replaced:

- ❏ Replace frayed or damaged electrical cords.
- ❏ Don't let cords get twisted, kinked, or crushed. Protect them from heat, chemicals, and oil. Protect extension cords on the floor with a cover designed for the purpose or use a board on each side of the cord (Fig. 13).

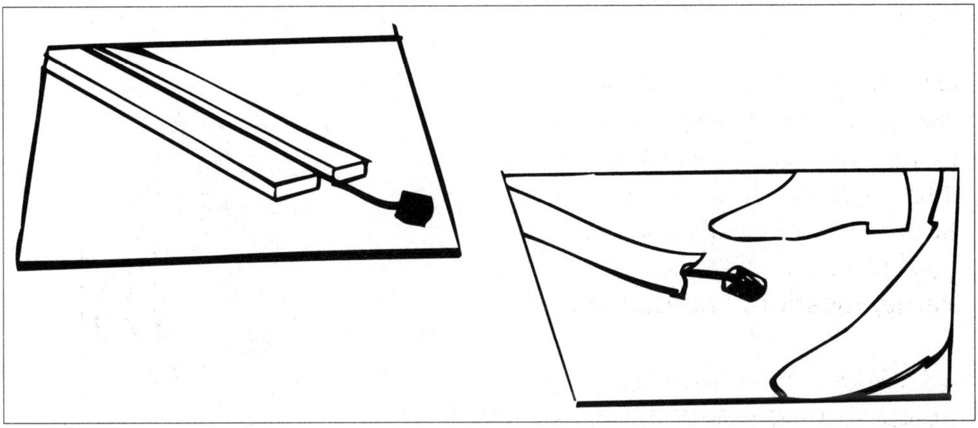

Fig. 13: Protect electrical cords on the floor

- ❑ Disconnect cords by pulling on the plug, not the cord.
- ❑ Don't overload a conductor. For power tools, use a heavy-duty extension cord that is rated for the load. A light duty cord will over-heat and may destroy the cord or cause a voltage drop that can harm equipment.
- ❑ Don't use a power tool if the grounding prong has been removed.
- ❑ If using a piece of equipment causes the power to go off, find out why. The tool may be defective and dangerous.
- ❑ If motor windings are dirty, have proper maintenance done.
- ❑ If equipment smokes, sparks, or gives a shock, **don't use it.**
- ❑ If a tool does not work consistently, get it repaired.
- ❑ Keep tools and equipment dry.
- ❑ Don't overload a circuit or piece of equipment by using too many electrical devices at the same time.

USE SAFETY FEATURES

Don't try to get around safety features. Safety features are there for a reason.

- ❑ If covers on switch boxes or other equipment must be removed temporarily, replace them as soon as possible. Exposed electrical contacts could cause electrical shock.
- ❑ If equipment has a safety switch that shuts it off automatically, use the reset button only once. If the safety switch turns off a second time, the equipment should be checked and repaired.

Fig. 14: Use an adapter properly attached

- ❏ Use a 3-prong grounded plug in a 3-hole outlet. Never cut off the third prong. If no 3-hole outlet is available, buy an adapter and connect it to the metal screw of the cover plate (Fig. 14).
- ❏ Never file or cut the wide prong of a polarized plug to fit it into an old outlet.
- ❏ Don't bypass fuses or circuit breakers or replace them with larger ones.
- ❏ Don't use an outlet if it may become damp—unless it is a GFCI (ground-fault circuit interrupter). This is an outlet that breaks the circuit if moisture or anything else causes an unbalanced circuit.

SAFETY TIPS

- ❏ Wiring may not have been installed according to code. Don't depend on its being correct. Don't assume by color that a wire is the neutral wire. It may be hot.
- ❏ Remove and replace motor starter covers carefully so that they don't touch a live contact in the box or pinch the insulation.
- ❏ If wires must be pulled out of a live box for reading amperage, pull them out carefully with your fingers. If you use a screwdriver to pry them out, you might cut through the insulation and touch the bare wire.
- ❏ Unless absolutely necessary, do not try to override interlocks that turn off the power when an electrical cabinet cover is removed. Sometimes it is necessary

to do this when measuring amperage. If so, be sure you know the manufacturer's recommended method of overriding the interlock, or ask an experienced technician how to do it.

- ❑ Don't store rubbish or rags near electrical equipment.
- ❑ Know where fire extinguishers are located. Find out if they can be used on electrical fires. Don't use water on an electrical fire.

WHAT TO DO IN CASE OF ELECTRICAL SHOCK

Every emergency is different. The general advice that follows may not fit every situation exactly. You must use your good judgment when you help someone who received an electric shock.

Don't become a victim yourself. Live electric current can flow from one body to another. Don't touch someone who is still in contact with electricity. If the victim is in contact with current, break the contact:

- ❑ Throw the switch for that circuit if you can.
- ❑ Use a piece of dry wood or other nonconductor to move the wire from the victim.
- ❑ If necessary cut the electrical cable to the victim using a tool with a wooden or plastic handle. Protect your eyes from the flash when the wires are cut.
- ❑ Drag or pull the victim away by using a nonconductor, such as a dry board, rope, blanket, coat, or belt.

Once the victim is free of the current, check for pulse, heart beat, breathing, or an obstruction in the throat. If the victim is not breathing, use CPR if you are trained. If not, try to find a qualified person immediately. Get help by calling 911 (or your local emergency number) or by having someone else call.

Loosen the victim's clothing to allow free breathing. Keep the victim lying down or resting. The heart may be very weak, and sudden activity could cause heart failure. Do not give the victim alcohol or medications. Keep the victim from getting too hot or too cold. Don't leave the victim until medical help arrives. There could be a relapse.

Your first responsibility is to keep from making yourself or someone else a victim of electricity. Keep a healthy respect for electricity and never just assume that conditions are safe. You may not have a second chance to guess right.

REVIEW

Answer every question before you check the answers in the back of the book.

1. Which of the following materials are good insulators? Which are good conductors?

   Water
   Dry air
   Glass
   Silver
   Rubber
   Aluminum
   Plastic
   Copper
   Human body

2. Why is a short circuit dangerous?

3. Name two devices for storing electricity.

4. What is a GFCI and what does it do?

5. How can you make sure that equipment connected by an electric cord stays safe to work on while you service it?

6. How can you make sure that equipment with permanent electrical connections stays safe to work on while you service it?

7. What are two dangers of using a light-duty extension cord for power tools?

8. What is your local emergency number?

9. Your buddy on a construction job has been knocked unconscious by an electrical shock and is still in contact with the live conductor. What are possible actions you could take immediately?

Give one possible answer to each of the following:

10. You turn off the power before working on a piece of equipment but get a shock anyway. Why?

11. You turn off the power to a piece of equipment and remove the electrical connections, but while you are working you suddenly get a shock. Why?

12. Dan once received a shock on a 110 volt circuit and was not seriously harmed, so he was relaxed about working with electricity. While working on another 110 volt circuit, he had a similar accident, received a shock, and died. Why?

Check your answers in the back of the book.

4

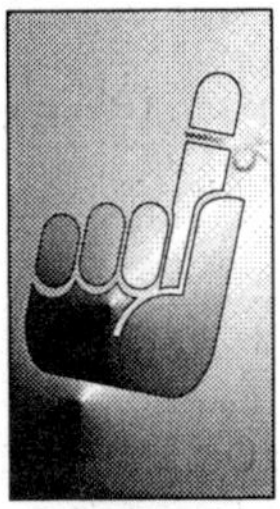

EQUIPMENT SAFETY

This chapter deals with safety with equipment such as furnaces, boilers, and other air conditioning equipment. It also covers safety with small tools. **Fan safety** is not covered in this chapter. Fans are so common in the work of the indoor environment technician that Chapter 5 is devoted to fan safety.

MOVING PARTS

When you work in an equipment room, there can be many exposed moving parts. Belt guards may have to be removed, exposing V-belts and drives. Pumps often have couplings with the motor shaft. There are common safety hazards from these moving parts:

- ❏ Machinery can start suddenly while a person is working on it.
- ❏ Hands can be injured from contact with moving parts.

Lock Out

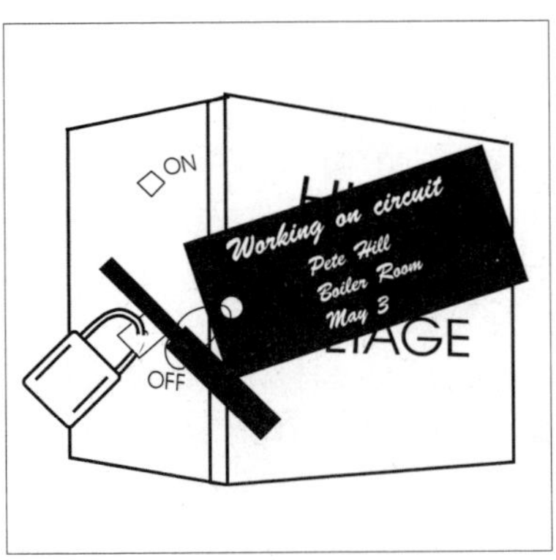

Fig. 1: Lock it out and tag it

The only sure way to prevent machinery from starting unexpectedly is to **lock it out** (Fig. 1):

- ❏ Disconnect the main switch to the machinery and put your padlock on it. You should be the only one with a key to the padlock. Attach a note or a tag to the lock stating who is working on the machine.

Test the power to the machinery to make sure the line is dead. A switch can be wired incorrectly and still send power to machinery even when it is in the OFF position. If the line is not dead, the equipment may start automatically. There should be a manual ON-OFF switch within reach. Try this switch to see if the equipment starts.

Block Out

Mechanical equipment may have moving parts that you should block out before you work on it. For example, blocks may be placed under raised dies, lifts, or other equipment that could fall or roll, causing injury. Spring-loaded devices should be released to prevent unexpected movement. Fan wheels should be blocked with a piece of wood even if the power is off.

Working Around Moving Parts

NEVER try to stop any moving part with your hands. A component may be coasting so slowly that it appears easy to stop. **You cannot stop it.** Your hands can be torn or crushed by being pulled into danger. Let any moving part stop completely before you touch it.

Never touch a moving pump coupling. They may have projections, such as set screws that can tear your hand. OSHA requires that couplings should not have projecting parts or should be covered by a smooth sleeve, but this may not always be the case for pumps already installed.

NEVER—under any circumstances—put your fingers under V-belts.

IMPROPER VENTING

Carbon monoxide (**CO**) is the incomplete product of combustion in fuel-burning equipment, such as furnaces and boilers. Carbon monoxide is a deadly gas and must be vented to the outside. The hot gases from the flames, including the carbon monoxide, travel through a **heat exchanger**. Water (for boilers) or air (for furnaces) is heated by contact with the outside of the heat exchanger. The hot gases from the heat exchanger travel to a vent pipe and are exhausted to the outside of the building.

If any carbon monoxide leaks into the building from a heat exchanger or vent, there is serious danger of illness or death. For this reason you should:

- ❑ Inspect heat exchangers regularly for cracks.
- ❑ Check vents for any improper connections and for holes. (Metal vent pipes can corrode because of acid condensing from the exhaust gases.)
- ❑ Never allow any part of a vent to be blocked off.

FLAME SAFETY

Gas-burning equipment is usually fueled by either natural gas or LPG (liquefied petroleum gas).

Natural gas has a distinctive odor, so it can usually be detected. It is lighter than air and dissipates with proper ventilation.

LPG (such as propane or butane) has an odor added so that it can be detected. It is very dangerous because it is heavier than air and settles on the floor or in other low areas. A fan may be needed to clear an area of LPG.

Gas-burning equipment has two main dangers from gas that is ignited by a spark or flame:

- An **explosion** due to leaks that allow gas to accumulate in the area.
- A **flashback** (fire spurting out of the opening) that could burn eyes and face when a burner is lighted.

If you suspect a gas leak, find the source of the leak and stop it. Dissipate any gas in the area.

- Do not hunt for gas leaks with a match. Finding the leak with a match can cause an explosion. Carry a small bottle of soapy water for detecting leaks. You can also use an instrument designed to detect gas leaks.
- If you smell gas, turn off the main gas valve to the building and open the area to outside air.
- If a pilot light goes out, turn off the gas valve to the equipment and wait at least 5 minutes before re-lighting. Ignoring this rule is a common cause of flashbacks that burn eyes and face.
- When lighting a pilot light or a burner, keep your face way from the opening. There can always be a flashback.

GUARDS

Guards are required over moving parts of equipment to prevent hands or clothes from being caught.

The technician often has to remove a guard. Always **replace a guard** before starting the machinery. If the guard must remain off while the machinery is running, put a barrier (such as a ladder) in front of it. This warns anyone that there is a danger. The barrier is a good reminder for yourself, because it is easy to forget the guard is off.

REFRIGERATION

Never open a refrigeration system unless you are a trained refrigeration service technician. There are various reasons for this:

- ❑ It is illegal to vent refrigerants to the atmosphere. They must be captured in tanks and turned in for recycling. Only a service technician certified to recover refrigerant may do this.
- ❑ There are dangers from burns, high pressures, and hazardous materials. Some CFC refrigerants can create a toxic gas in the presence of flames and some newer refrigerants in centrifugal chillers are hazardous.

HAND TOOLS

- ❑ Replace **defective tools** immediately. This seems so simple that it should not need to be mentioned. Yet most of us are guilty of unsafe work practices because a defective tool requires a little extra time to replace. A hammer with a loose handle is easy to ignore until it flies off and injures someone—or breaks a window.
- ❑ **Dull tools** are a common source of accidents. They require more pressure to use so your hand is more likely to slip.

Fig. 2: Mushroom head

- ❑ A **mushroom head** (Fig. 2) forms on a chisel or other tool that is pounded on constantly. The head curls over and the steel becomes brittle. Eventually, chips will fly off with such force that they can penetrate the skin or eyes. Grind off the first sign of mushroom heads.

- **Never carry tools in your pocket**. They can gouge you or another person. They can cause greater injury if you fall. They can hang up on an obstruction at the worst possible moment. Carry hand tools in a pouch.
- Always use the **correct tool** for a task.

Screwdrivers

For construction work, battery powered screwdrivers are common. However, technicians commonly use hand screwdrivers, and they often cause accidents:

- A dull or worn tip of a standard screwdriver can slip out of the screw slot. This is annoying and can cause your hand to slip into machinery or sharp objects. Occasionally touch up screwdrivers on a grinding wheel. The characteristics of a properly sharpened screwdriver are shown in Fig. 3.
- Avoid holding small objects in your hand when working on them with a screwdriver. If the screwdriver slips, you will gouge your hand.

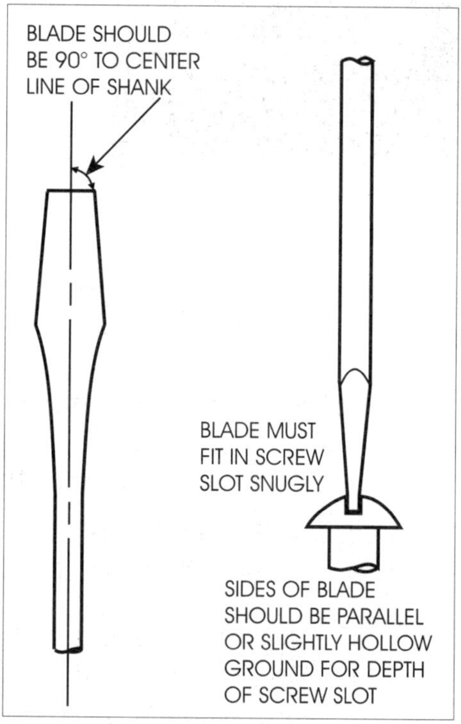

Fig. 3: Properly sharpened screwdriver blade

- Never pound on a screwdriver. It can damage the handle, bend the shaft, and damage the tip.
- Use only screwdrivers with plastic handles. The handles are almost indestructible, and they protect you from electric shock.

Wrenches

Wrenches can slip and cause you to gash your hand against something. Protect yourself from these dangers:

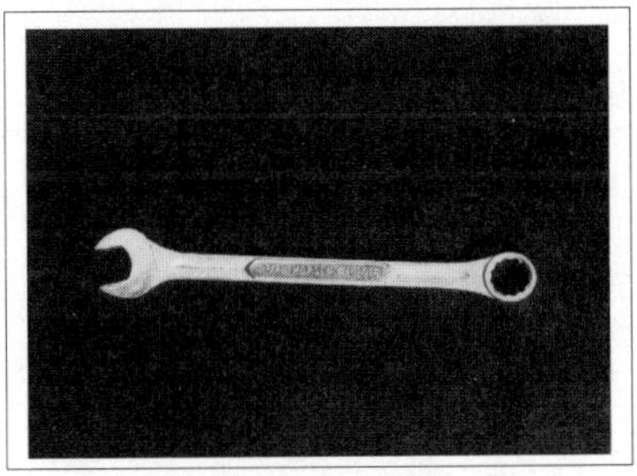

Fig. 4: Box end wrench

- Use the proper size wrench. Use a **box end wrench** (Fig. 4) when it is practical.
- Always try to pull towards you instead of pushing on a wrench. If the wrench slips, you are less likely to hit your hand on some object if you are pulling rather than pushing. If you must push on a wrench, do it with care and try to do it in a direction where your hand will not hit a sharp or hard object.

When using a crescent wrench (adjustable end wrench):

- Adjust it so it fits the bolt or nut tightly.
- Locate the wrench so that the pressure is on the fixed jaw of the wrench, not on the adjustable jaw.

If a nut is frozen and will not move, use penetrating oil and let it stand for a while. Other measures are tapping the side of the nut lightly, or applying heat if practical. The joking advice, "Don't force it—get a bigger handle" is too often followed. Never use a pipe on the end of a wrench to get greater leverage. You can damage the wrench, and if the bolt or nut is frozen tight, the bolt can twist and break off.

Large Drill Motors

Be aware that the torque (the twisting force) on a large drill motor is very great. If a twist drill hangs up in a hole—as they often do—the drill motor is hard to stop. If you are not braced properly, the force can throw you off balance. This can cause a bad fall, especially for someone working on a ladder or scaffold.

A twist drill usually hangs up at the point when the drill is breaking through the metal. If you are attentive, you can tell by the sound when the drill is about to break through. Let up on the pressure on the drill at this time and hanging up can usually be avoided.

Powder Actuated Tools

A **powder actuated tool** (Fig. 5) uses a powder charge to shoot a fastener into concrete. Many areas require that the operator have a license that can only be obtained by completing a safety training course. Operators and those working near them must follow these general safety rules:

COURTESY HILTI, INC.

Fig. 5: Powder actuated tool

- ❑ **Always wear safety goggles** when using or working near powder actuated tools.
- ❑ When the fastener is in the tool, it is like a loaded gun. Many workers have been killed or injured from being struck by a discharged fastener.
- ❑ A safety device prevents these tools from being fired unless they are pressed against something like a wall. However, never depend on this safety feature. Treat the tool like a loaded gun and never point it at

anyone, even if you think that the powder charge is not in the tool.

- ❑ If the fastener hits a piece of steel hidden in the concrete wall, the fastener is like a ricocheting bullet. The best protection against this is to always wear safety goggles and a hard hat.

Pneumatic Lines

Pneumatic lines are rubber hoses used to deliver pressurized air to operate power tools. Pneumatic hose lines can be dangerous:

- ❑ Most air lines have a quick disconnect fitting that automatically shuts off the air when the hose is disconnected from a power tool. However, be aware that an air hose that is disconnected without being shut off can whip around dangerously.
- ❑ Do not use compressed air to blow off your clothes. It can cause air blisters under your skin. More important, compressed air contacting oil can create spontaneous combustion. If your clothes happen to be oily, they can catch on fire.

REVIEW

1. What are three problems to look for when inspecting a heat exchanger and vent?
2. What type of fuel gas can settle on the floor or other low areas?
3. What type of fuel gas has a distinctive odor?
4. If you suspect a gas leak, how do you look for the leak?

5. If a pilot light goes out, how long should you wait before relighting it?

6. Should you pull towards you or push away when using a wrench?

7. If you have a choice, should you use a box end wrench or a crescent wrench?

Check your answers in the back of the book.

5 FAN SAFETY

The fan provides the energy that moves air through the ducts to provide comfort to the spaces. Indoor environment technicians work on fans in many ways. They install fans, measure fan performance, and adjust fan speed.

Fan terms vary in different parts of the country. The terms used in this chapter are defined as follows:

Fig. 1: Fan plenum in the equipment room

Fig. 2: V-belt drive

- **Equipment room**: A room where all the central HVAC equipment is installed (piping, pumps, fan plenum, etc.)
- **Fan plenum**: A large metal enclosure (Fig. 1) in the equipment room. It is airtight in order to force the air path through the filters, coils, and fan. The fan plenum is often large enough for a person to enter and stand upright in order to service the fan, coils, and filters.
- **V-belt drive** (Fig. 2): The system that transmits power from the motor to the fan. It has three parts:

  Drive sheave: The sheave on the motor.

Driven sheave: The sheave on the fan shaft.

V-belt: The belt that fits on the sheaves and transmits power from the motor to the fan. A small fan may have only one V-belt. Large fans have several V-belts.

MOVING PARTS

Safety rules on working with moving parts are covered in Chapter 4, Equipment Safety. These rules are just as important for fan safety.

Lock Out

The only sure way to prevent a fan from starting unexpectedly is to **lock it out**:

- ❑ Disconnect the main switch to the machinery and put your padlock on it. You should be the only one with a key to the padlock. Attach a note or a tag to the lock with your name and the date.
- ❑ Test the power to the fan motor to make sure the line is dead. A switch can be wired incorrectly and still send power to machinery even when it is in the OFF position. If the line is not dead, the fan may start automatically. There should be a manual ON-OFF switch within reach of the fan motor. Try this switch to see if the equipment starts.

Block Out

A worker who was changing the V-belts on an exhaust fan de-energized the fan before starting work and locked out the switch. However he did not block the blades of the fan. The suction in the ductwork turned the fan blades, and his hand was caught in the V-belt drive.

This example shows the importance of blocking out a fan or any mechanical part that could start moving even with the power off.

V-belt Drives

The advice given for all moving parts in Chapter 4 is especially important when working with fans. **Never** try to stop a V-belt drive with your hands. Let it stop completely before you touch it. A fan sheave may be coasting so slowly that it appears easy to stop. **You cannot stop it**. The combined momentum of the fan, motor, and drives is too strong. Your hands can be torn or crushed by being pulled into danger. Some sheaves may have protruding set screws. These will tear your hand even it there is only a slight rotation.

Never–under any circumstances—put your fingers under V-belts. When working on a V-belt drive, even with the equipment locked out, it is easy to move the fan sheave slightly and have your fingers drawn in between the belts and the sheave. If the equipment somehow starts and your fingers are under the belts, you will probably lose your fingers.

Belt Guards

Fig. 3: A guard must be in place over the V-belt drive

Fans must have guards (Fig. 3) over the V-belt drive to prevent hands or clothes from being caught. You often have to remove a guard. Always replace the guard before starting the fan. If the guard must remain off while the machinery is running, put a barrier (such as a ladder) in front of it. This warns anyone that there is danger. The barrier also reminds you of the danger, because it is easy to forget the guard is off. The barrier is especially important in a fan plenum. This is usually a dimly lit place where it is not easy to see the exposed machinery.

Changing the V-belt Drive

Often the technician has to change the fan output. In order to change the fan output in CFM (cubic feet per minute), the fan RPM (revolutions per minute) must be changed. The usual way to do this is to remove the V-belts and change the motor or fan sheave to one of a different diameter. This requires the following safety procedures:

- ❑ Do not change the drive system unless:
    - ❑ You know how to calculate the new RPM that results from the change.
    - ❑ You know the RPM limit of the fan. Speeds over the designed limit of the fan can damage the wheel.
- ❑ Be sure the fan motor switch is locked out. Test the line to be sure the power is off.
- ❑ NEVER use a steel hammer directly on the sheave to remove or install it. You will damage the shaft or the sheave. If they do not move readily, do some of the following:
    - ❑ Some have provisions for using bolts in the bushing to push the sheave off the shaft.
    - ❑ If there are no bolts in the bushing, use a **wheel puller** (Fig. 4).
    - ❑ If the sheave cannot be moved, apply penetrating oil and let it soak in for a while.

Fig. 4: Wheel puller

  - ❑ If you must pound to move the sheave or bushing, do it as a last resort. Hold a piece of wood firmly against the hub and pound on the wood. Alternate from one side of the hub to the other. Be aware that the pounding force on any part but the hub will break or bend the sheave.

- ❑ Use a cord or a straight edge to be sure that the sheaves are properly aligned with each other and are not at an angle. Align the sheaves if necessary.

IN THE FAN PLENUM

Fan plenums are usually dimly lit and have very little extra room. In addition, they have moving parts such as V-belt drives. They require special safety precautions when working in them:

- ❑ Let someone know you are entering a plenum, in case you get trapped.
- ❑ Do not wear loose clothing or long sleeves that can catch on moving parts.
- ❑ If you have long hair, be sure that it is tied up or under a cap.
- ❑ Carry a flashlight. You will need it in dimly lit areas.
- ❑ There is little head room in a fan plenum. When you enter, use a flashlight to look for all the possible places where you could hit your head. Electric lights in fan plenums may have waterproof covers which are often very low and a hazard to the head.
- ❑ Carry a cordless phone, especially if you are working alone, so that you can call for help if you are injured or trapped in a fan plenum.

Always open the fan plenum doors carefully. Excess pressure can blow the doors open violently on the fan discharge side.

You can be trapped in a fan plenum. If there is no shut-off switch in the plenum and the outside air and return air dampers close, the fan will create a negative pressure in the plenum. This pressure can easily be strong enough to prevent you from pushing open the plenum door.

Before working in a fan plenum, remove all debris and stored material. The plenum is sometimes a favorite place to store material. Anything stored in the plenum can contaminate the conditioned air. It also increases your chance of tripping.

Make sure you have all the tools you need before you enter the plenum. Once in an awkward position, it is so much trouble to get out and get the proper tool that you will be tempted to do foolish things such as using a knife for a screwdriver or using a crescent wrench for a hammer. These lead to injuries, damaged tools, and a poorly done job.

Electrical Safety in the Fan Plenum

Review Chapter 3, Electrical Safety. Follow all electrical safety rules carefully. In the fan plenum you are working in close quarters and can easily be grounded against metal. Codes require that the fan plenum must have a disconnect switch within sight of the fan motor.

Avoid using extension cords in the fan plenum. Use battery-powered tools instead. If you must use cords, be very careful that they remain clear of any moving machinery.

REVIEW

1. Under what conditions can you stop a V-belt with your hands?

2. If a fan motor is running without the guard in place, what safety measure should you take?

3. What is the first action to take when changing the V-belt drive system?

4. What items are likely to be included in the fan plenum?

5. Why should you never put your fingers under V-belts?

6. If you must work by yourself in a fan plenum, what safety measure should you take?

7. Label the drive sheave and the driven sheave.

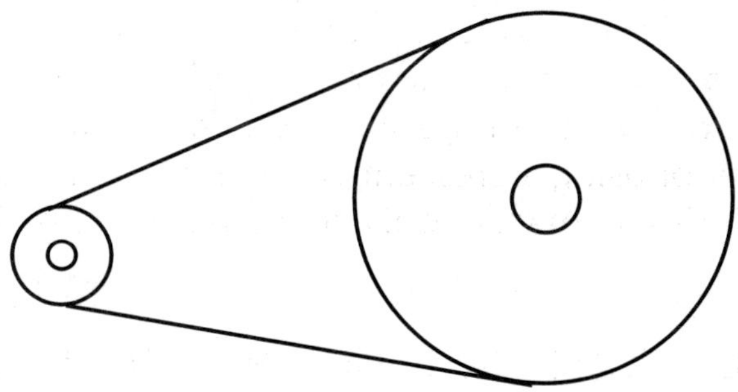

Check your answers in the back of the book.

6 PIPING SAFETY

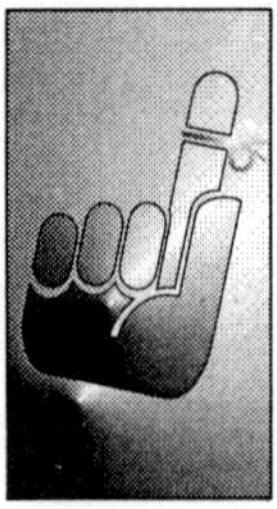

WORKING WITH THE HYDRONIC SYSTEM

The indoor environment technician is responsible for the performance of the entire HVAC system. This includes the air, electrical, refrigeration, and **hydronic systems**. The **hydronic system** in HVAC work is the **piping system** which conveys heat to or from the space by means of chilled water, hot water, condensing water, or steam.

The technician usually works on the piping systems in order to test for water temperature and flow rates. Sometimes the technician must install an automatic control system component or clean a strainer. The technician may recommend retrofit of a piping system, but this work is usually performed by another craft.

Many dangers are associated with working on the hydronic system. Learn to avoid them and keep yourself safe.

Identify a piping system before you work on it. Hot water and steam can burn you badly. Usually the piping in a system is labeled. If it is not labeled, touch the pipe to see if it is hot. Insulation on a pipe is a warning that the fluid is very cold or very hot.

Observe carefully all the safety rules for working with electrical systems. These are covered in Chapter 3. When working with hydronics, you are in contact with water and metal pipes. This means that you will probably be very well grounded, which makes electrical shock especially dangerous. Lock out and tag any electrical equipment you will work with.

Always wear safety goggles when working with piping. Hot water, steam, dirt, and chemicals are all hazards to your eyes.

VALVES

Fig. 1: Pressure relief valve

Pressure relief valves (Fig. 1) limit the pressure on the system. They are safety devices that release pressure if it builds up too much. Locate the pressure relief valves on a system before you work on it.

Never disable a pressure relief valve. Doing so could damage the system. Even more important, someone who opens the system could be injured if the pressure has built up to dangerous levels.

PT plugs (pressure/temperature taps) are fittings that allow a thermometer or a pressure measuring device to be inserted in the pipe (Fig. 2). They are designed to open when a device is inserted and to close automatically when it is removed. Sometimes a thermometer is left in a PT plug to keep a continual reading on the temperature (Fig. 2). If it has been in for a long time, the PT plug may not close. For this reason, never stand in front of a PT plug when removing a thermometer or other device.

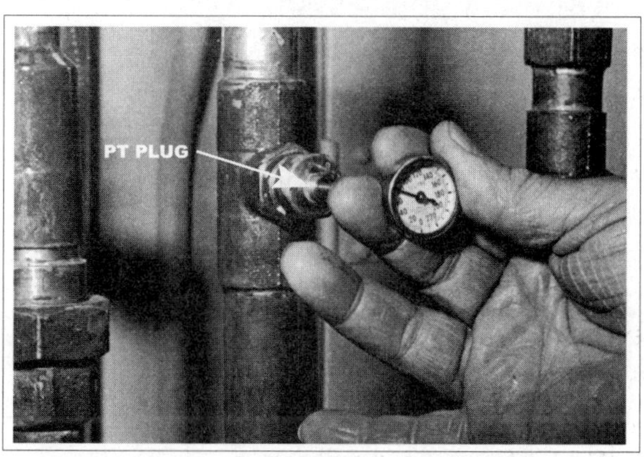

Fig. 2: Using a thermometer in a PT plug

Close manual valves slowly. Closing one too fast may cause a physical shock to the system that can break cast iron fittings and also knock pipe lines off their hangers.

Packing glands on valve stems may leak when the valves are closed after having been open for a long time.

ENTERING A SYSTEM

Entering a system means opening some part of it. This exposes the piping system to the atmosphere. Opening a valve to the atmosphere is entering the system. The system may have to be entered to replace a component or to install a new one.

An **open system** is one which has the water normally exposed to the atmosphere. For example, an open system may have a cooling tower which allows outside air to pass over the water from the hydronic system. A **closed system** is not exposed to the atmosphere until it is entered.

When a system is opened, there is always some danger of damage to the building, the system, or the technician. The greatest dangers involved in entering a system are these:

- ❑ Burns from steam or hot water
- ❑ Exposure to contaminated water or hazardous chemicals used for water treatment
- ❑ Damage to the building from water or steam

Water spills can happen easily when working on piping systems. They are hazardous in many ways:

- ❑ They can cause slips and falls.
- ❑ Wet floors and shoes increase the severity of any electrical shock.
- ❑ Hot water or steam can cause severe burns.

- Spills can damage the building. Even small amounts can cause stains.
- Water from open systems can spread disease. This is especially true of water from cooling towers and coil pans.

You may have to drain the system (or a portion of it that can be isolated) before you work on it. Take proper precautions to avoid damage to the building and danger to yourself when you enter a system. Follow these procedures to protect yourself and the building:

- Wear eye protection and gloves. Water and chemicals in a system should never touch your skin.
- Find out what chemicals are being used for water treatment, and find out how hazardous they are.
- Relieve the pressure on the system before opening it. Do this by cracking a valve or by activating a pressure relief valve.
- Be sure the floor drain is not plugged before draining the system.
- Know where the shut-off valves are located. You do not want to hunt for them in an emergency.
- Attach a hose to the system at a low point and drain the water into a floor drain.
- Don't release very hot water into a floor drain. Dilute it with cold water. Floor drains are not designed to carry hot water.
- Be aware that gate valves may not shut off completely.

The most common reason a technician has for entering a hydronic system is to clean a strainer. To do this, follow these steps:

- ❏ Isolate the strainer so it can be drained without emptying the entire system.
- ❏ Drain down the piping near the strainer.
- ❏ Remove the end cover from the strainer.
- ❏ Pull out the sleeve and clean it.
- ❏ Replace the cleaned sleeve and close up the system.
- ❏ Open the isolation valves while venting as much entrained air as possible.

REFRIGERATION SYSTEMS

It is illegal to open a **refrigeration system** unless you are certified to capture and recycle refrigerants. In addition, it is dangerous to open a system unless you are well trained in this type of work. At certain places in the system, refrigerants can be hot and under high pressure. Some refrigerants are toxic and some become highly toxic in the presence of a flame.

PUMPS

The technician sometimes has to work on pumps to change flow rates. This usually means removing the pump cover to change the impeller diameter. This means that you are opening the system. Observe all safety precautions related to water spills and entering a system. Also observe the following:

- ❏ Follow lock-out procedures to avoid unexpected start-up.
- ❏ Carefully observe electrical safety rules. The presence of metal pipe and water can make any electrical shock very severe.

- ❏ Do not let your shoes get wet. It will increase the hazards of electrical shock here and any other place you work.

When lubricating a pump or its motor, carefully follow the manufacturer's instructions to prevent damage to the bearings and seals.

REVIEW

1. What is the purpose of a PT plug?
2. What type of valve serves as a safety device?
3. Should manual valves be closed quickly or slowly? Why?
4. What is the difference between an open system and a closed system?
5. What danger is associated with open systems?
6. What action should you take before opening a system?

Check your answers in the back of the book.

7 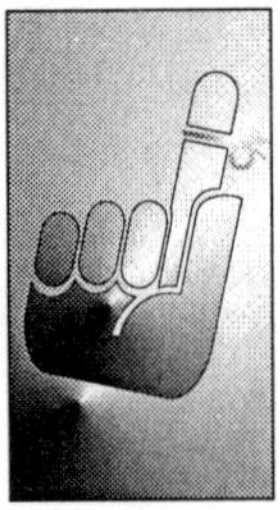 SHEET METAL SHOP SAFETY

Some indoor environment technicians are sheet metal workers who fabricate air conditioning duct in the shop. These technicians must be aware of the hazards in the sheet metal shop.

GENERAL SAFETY

- ❑ When grinding or working near grinding, wear proper eye protection.
- ❑ When grinding or disk sanding, control the direction that the sparks are flying. Direct them away from people and fire hazards.
- ❑ Wear ankle-high leather shoes. Athletic shoes offer little protection from cuts or crushing.
- ❑ Watch where you are walking. There are many sharp objects that can cut and many things to stumble over.
- ❑ The edges of sheet metal sheets can easily slice your hands. Never let the metal edge slide against your palm.
- ❑ Cut off sharp points and corners from parts of sheets before you store them.
- ❑ Pick up scrap metal and other debris continually. It is a frequent cause of tripping.
- ❑ Get help in lifting and moving heavy loads. Use available hoisting and moving equipment.
- ❑ Never remove a safety guard on any piece of machinery unless it is required in order to adjust the

machine. Remove the guard only when the machine is not running.

- ❑ Do not start a machine unless all safety guards are in place.
- ❑ Alcohol or drugs are not tolerated. Refuse to work with anyone under their influence.

SHEARS

Both the small foot-operated squaring shear and the power shear can cut off fingers swiftly. Use them safely:

- ❑ Never remove any safety guards.
- ❑ Never put your fingers under the shear blades.
- ❑ Do not try to hold small pieces with your fingers.
- ❑ Never place your fingers where you cannot see your fingertips.
- ❑ Do not allow anyone behind the shear when you are operating it. They can be hit by the sheet metal or by parts of the back gage.

HAND BRAKES

- ❑ If you are working on the eight-foot brake with a partner, always be aware of what the other person is doing. This is especially true if the other person must reach into the brake.
- ❑ Be sure that no one is standing where they can be hit by the swinging parts of the brake.

PRESS BRAKES

- ❑ Never remove any safety guards from a press brake.

- ❑ Never put your fingers between the dies of a press brake.
- ❑ If you are working with a partner on the press brake, the two of you must watch out for each other. Always be aware of what your partner is doing. No surprises! Work as a team and keep each other informed about what will happen next.
- ❑ Lower the upper die at a moderate rate to make the bend. Fast action can snap the metal upward and injure your hands or those of your partner.
- ❑ Press brake dies can be very heavy. Many shops have lifting devices and rolling tables to move the dies in place. If you must move the dies by hand, get someone to help you.

FORMING ROLLS

- ❑ Do not wear gloves or loose clothing when operating a power forming roll. It is easy to catch glove tips in the rolls so that your hands are pulled in and crushed.

COILS OF METAL

Large coils of sheet metal are used in machines such a coil line for fabricating straight duct. A coil of metal weighs several tons. If it gets out of control, it can easily crush someone.

- ❑ Have help in moving and placing coils.
- ❑ Move the coils only by means of the hoisting equipment provided for it.
- ❑ Coils are held by metal bands. These are usually under considerable pressure and snap apart with great force when cut. Protect your face and warn others nearby.

SOLDERING

- ❑ Many soldering fluxes are corrosive. They can blind you or burn your skin.
- ❑ If you get any flux in your eyes, wash them out with cold water and see a doctor immediately.
- ❑ If flux splashes on your skin or on your clothes, wash it off with cold water immediately.
- ❑ Some portable firepots use flammable fuel, such as white gasoline. Treat it with extreme care.
- ❑ Keep your face away from a firepot when lighting it. There is always a danger of a flash.
- ❑ Solder contains lead. Wash your hands after handling solder or other lead products. The soldering area should be well-ventilated.

WELDING

Welding of any kind requires instruction in how to use the equipment safely. Do not use any welding equipment unless you have been instructed in its safe use.

In the shop and on the job, you may be working around a welding process, or you may have to assist a welder. Practice the following safety precautions:

- ❑ All welding processes require eye protection with special shaded lenses. When welding or working around welders, be sure to have the proper type of eye protection.
- ❑ The rays from arc welding will burn your eyes if they are not protected with the proper shade of eye protection. Do not look at the arc without glasses. If you are walking by an arc welding process, turn your eyes or shield them with your hand.

- In the shop, a screen should be set up around the welding area to prevent the flash from burning the eyes of others nearby.
- Oxygen and acetylene tanks must be kept upright. They should be chained or tied with a rope to something solid to guarantee that they cannot fall over:
    - Oxygen is under high pressure. If the valve is broken off, it is as dangerous as a live rocket.
    - Acetylene is highly flammable. If the valve is broken off, there is danger of a violent fire.
- Oil exposed to high pressure oxygen can result in spontaneous combustion and a violent fire. Never oil the oxygen gages or allow oil near the oxygen tanks.
- Never wear oily clothes or oily gloves when working with oxygen because they are so flammable.
- The fumes from welding galvanized iron can make you ill. Use some method of ventilation so that you do not breathe these fumes.

REVIEW

List at least two safety rules when working with each of the following:

- Shears
- Hand brakes
- Press brakes
- Coils of metal
- Soldering
- Welding

8 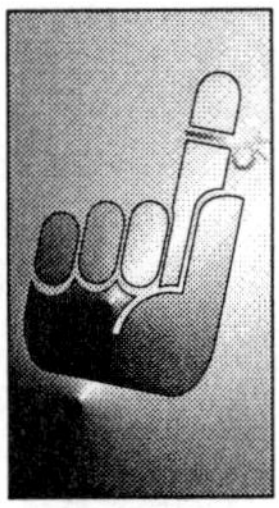 CONSTRUCTION SAFETY

PREPARING FOR EMERGENCIES

Accidents can happen on a construction site. An indoor environment technician often works on the HVAC system of a building that is under construction or is being renovated. This creates many special hazards, especially for falls.

Every job site must be prepared for emergencies. Then, when accidents happen, there will be less confusion and more effective emergency care. OSHA requires the following for every job site:

- ❑ **First aid kit**
  A fully stocked first aid kit must be located in a central place on the job. On a construction job it is often located in the office of the job superintendent. Often job superintendents carry a kit in their truck.
- ❑ **Someone trained in first aid**
  The employer must have someone on the job who is trained in first aid. Several people on a job crew may have first aid training. On a construction site, the general contractor usually makes sure there is a trained first aid person for the entire job.
- ❑ **Emergency phone numbers**
  Emergency phone numbers for fire, police, and ambulance must be posted.

RESCUES

An accident victim may need to be rescued. If you have to rescue someone, **don't become a victim yourself.** Too often one person trying to rescue another becomes a second victim. For example, a rescuer grabs a buddy who is in contact with a live electrical circuit and the rescuer is also electrocuted. Or a rescuer goes into a tank to bring out a worker overcome by fumes and the rescuer also dies of the fumes.

When an accident happens, the first impulse is to rush in to help the victim. The first rule of rescue is this:

Don't act on impulse.

Think carefully about what you are going to do. Don't attempt a rescue unless the victim's life is in immediate danger or there is danger of further injury. It may be better to wait for professionals to take over.

If rescue is a necessity, avoid doing it alone. Have some back-up to help you in case you get into trouble.

PROVIDING AID

Knowing when to give medical aid requires judgment. The wrong treatment may cause further injuries.

For serious injuries, the basic guideline for first aid is to treat only life-threatening injuries. Usually this means that you should treat excessive bleeding or stopped breathing. Of course, this rule must be tempered with good judgment—there are exceptions. For example, if someone has been splashed with acid, you wash it off. If someone is burned by hot asphalt, you apply cold water.

Minor injuries may receive first aid. Clean and bandage small cuts that do not need emergency care.

When there is an accident, follow these guidelines:

- ❑ **Call for professional help immediately.**
  Call 911 or whatever your local emergency number is. The dispatcher will send an ambulance and paramedics if needed. Police may be needed. The caller must be calm enough to provide the following details:

  The location, including the nearest cross streets

  The nature of the accident

  Whether a rescue is needed

  The number of victims

  Approximate condition of the victims and what is being done for them

  Name and phone number of caller

  The caller should wait on the line until the person receiving the call is finished asking questions. **Do not hang up first**.

- ❑ **Do only what is necessary**.
  Treat excessive bleeding, stopped breathing, or other conditions that require immediate attention. Do not move the victim, unless it is really necessary. Moving the victim can make back injuries and broken bones worse.

- ❑ **Treat for shock.**
  Keep the victim warm and lying quietly.

- ❑ **Provide help and encouragement**.
  A major part of your job is to keep the victim quiet and to avoid panic. Assure the victim that help is on the way. Fear or panic are harmful for the victim.

PREVENTING ACCIDENTS

It is much better to prevent accidents than to deal with the results. Be aware of potential hazards. For example, you need to know the dangers of working in a confined space and take proper measures to ensure safety. These were covered in Chapter 2, Hazardous Materials. **Warning lines** are used on the ground to keep passersby from walking into dangerous areas. For example, warning lines are put up around areas where material is being stocked or hoisted or where a roofer's kettle is heating hot asphalt. Many other hazardous conditions face you on a construction site.

Fall Protection

Fall protection is any structure or equipment that prevents a fall. This includes warning lines, guardrails, catch platforms, and fall restraint gear. Job conditions and OSHA regulations determine what kind of fall protection is needed.

On a construction site, a very common accident is falling through holes in roofs and floors. All openings must have a guardrail around them or a cover over them:

- ❑ A **cover** must be strong enough to hold a person's weight and must be fastened down. If the cover is not fastened, a worker can easily pick up what appears to be a loose piece of plywood. In picking it up, it is natural to step forward. As a result, the person steps into the hole.
- ❑ **Guardrails** (Fig. 1) must be used around stair openings and may be used instead of covers around other openings. Guardrails must be strong enough to withstand the weight of a person leaning against them.

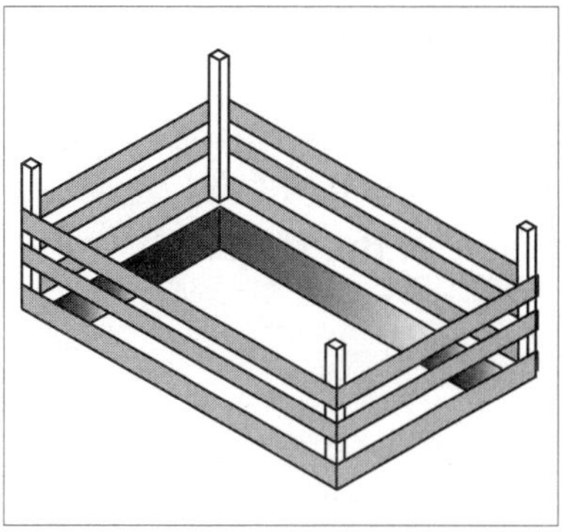

Fig. 1: Guardrails around an opening

Fig. 2: Fall arrest gear stops a worker who falls

Personal **fall arrest** equipment stops a worker who has fallen from a roof or other high point. It consists of a **safety harness** (not a belt), a **lanyard**, and a **lifeline** (Fig. 2):

- ❏ The worker wears the **safety harness**.
- ❏ The **lanyard** is a short connector between the safety harness and the lifeline. It is designed to absorb the shock at the end of a fall and greatly reduce the shock to the worker. The lanyard must be short enough so that a fall can be no more than six feet.
- ❏ The **lifeline** is anchored to a substantial object that will resist the shock of the fall and hold the weight of the worker.

Personal **fall restraint** equipment is designed to **prevent** a fall, not to **stop** a fall. It keeps a worker from reaching a roof edge or other hazardous area. It usually consists of a safety harness or belt, a lanyard, and a lifeline.

The employer is required to provide fall protection equipment and to train workers how to use it properly.

REVIEW

1. List six items of information you should give when you call for help for a serious job site accident.
2. What is the difference between **fall arrest** equipment and **fall restraint** equipment?

3. What are the three basic parts of a fall arrest system?

4. You discover a fellow worker on the ground at the base of a scaffold. He is not conscious but is breathing. He is not bleeding. List the steps you would take in the order you would do them.

Check your answers in the back of the book.

9 LADDER SAFETY

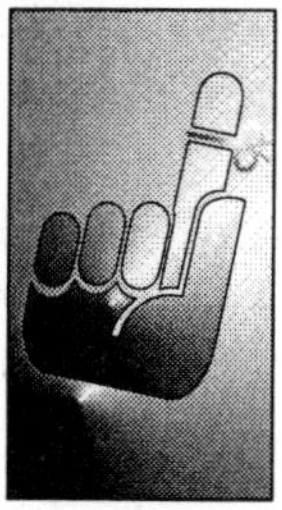

AVOIDING ACCIDENTS

Ladders are used almost daily on the job. They are a very frequent cause of accidents:

- Workers fall from the ladder.
- The foot of the ladder can slip out.
- The ladder can tip over.

Any ladder accident that happens to you means pain and hardship for you and your family.

Use a ladder whenever it is needed. It is a temptation to use something else if ladder is not easily on hand. Never use a makeshift device, such as an upturned bucket or an oil drum. You are greatly increasing your chances of a fall.

Ladders on the job have a tendency to wander. Some other worker can borrow it "for just a minute." Then someone else may take it from there. Soon you have lost track of it. You are tempted to use a makeshift if the ladder is not where it should be. All ladders on the job should be painted with the company name. You should know where they are at all times.

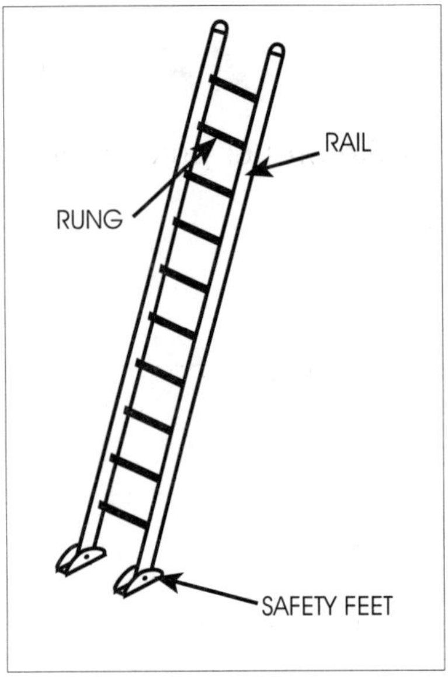

Fig. 1: Straight ladder

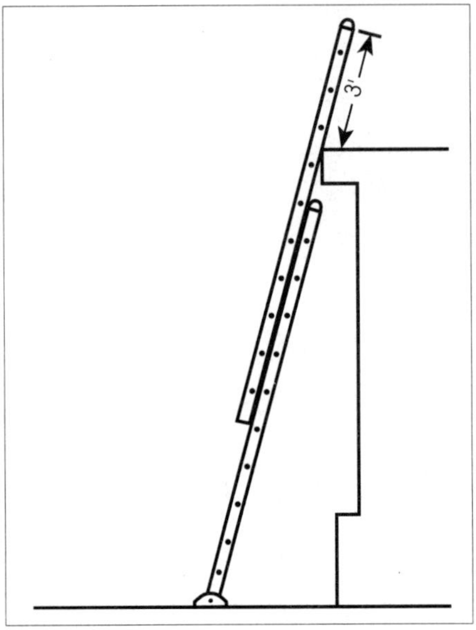

Fig. 2: Extension ladder

Common types of ladders used on the job are the straight ladder (Fig. 1), the extension ladder (Fig. 2), and the step ladder (Fig. 3). Figure 1 names the important parts of the ladder.

Damaged Ladders

Ladders must be in good condition. A broken rung can cause a fall. Slits in the rail can result in a broken ladder and a fall. Return a ladder to the shop if any part is broken or cracked or not working properly. Never use a ladder that is painted. The paint can hide splits and cracks.

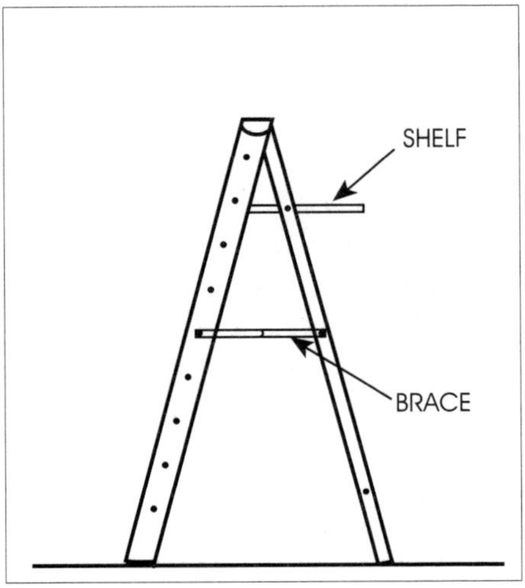

Fig. 3: Step ladder

You are not doing your employer a favor by leaving a damaged ladder on the job. An accident caused by a damaged ladder can lead to fines from OSHA. Any accidents on the job lead to increased Workers' Compensation Insurance costs.

SETTING UP LADDERS

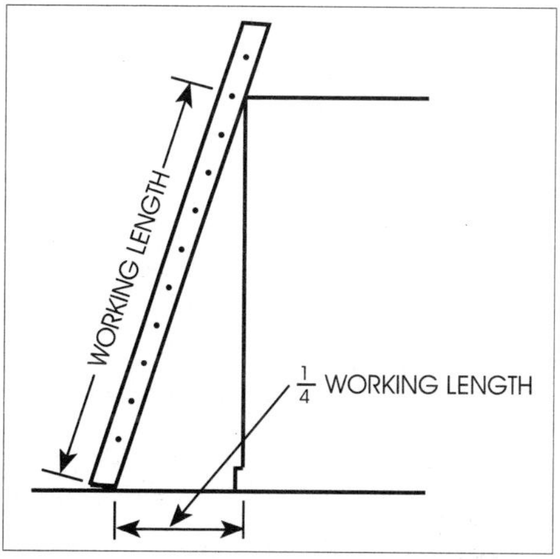

Fig. 4: The distance of the base from the wall should be ¼ of the working length

To set up a straight ladder or extension ladder, use the 4 to 1 rule (Fig. 4). This means that the bottom of the ladder should be 1 foot away from the wall for every 4 feet of vertical height. For example, if 10 feet of the ladder are leaning against a wall, the bottom of the ladder must be 2½ feet away from the wall. The rungs are about one foot apart, so it is easy to figure the working length of the ladder.

At an angle of 4 to 1 the ladder is less likely to tip away from the wall, and the feet are not likely to slip out from the wall.

A good rule-of-thumb is to stand with your feet at the foot of the ladder and extend your arms straight out in front of you. If the ladder is at the proper angle, your finger tips should just touch the ladder rails.

Position a ladder safely:

- ❑ Do not lean a ladder against a door.
- ❑ Do not position a ladder where it can be struck by an opening door or window.
- ❑ Keep at least ten feet away from electrical wires.
- ❑ Keep clear of other construction activities, such as hoisting.

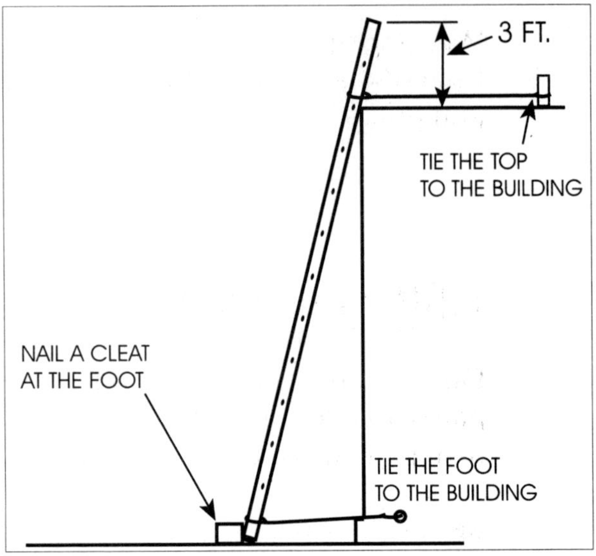

Fig. 5: Three alternate ways to secure a ladder

Soft or uneven ground is a common cause of tipping. If the ground is soft, a leg can sink when weight is placed on the ladder. The ladder can tip sideways. Place the ladder on a level board if the ground is unstable or uneven.

Always use some method to keep the base of a ladder from slipping out away from the building. Many ladders have rubber **safety feet** (Fig. 1) to prevent slipping. They are pivoted so that they are always flat on the ground. However, do not depend only on the safety feet to protect you. The ladder should also be tied off or should have the feet blocked. Use the method that fits the conditions:

- ❑ Tie the foot of the ladder to the building (Fig. 5).
- ❑ Tie the top of the ladder to the building (Fig. 5).
- ❑ Nail a cleat at the foot of the ladder (Fig. 5).

OSHA requires that a ladder must extend at least 3 feet above the upper level. If this is not possible, the top of the ladder must be secured from slipping, and hand holds (such as a grab rail) must be provided at the top. This is to prevent falls at the top of the ladder. The area around the top of the ladder must be clear of materials and debris. Someone should be stationed at the base of the ladder until it is tied off.

Warning lines or a barricade should be placed around the base of the ladder if it is in an area used by people or equipment.

USING LADDERS

Do not carry anything up a ladder. You must have both hands free for climbing the ladder. Hoist equipment and materials by rope.

Be careful when you have to work from a ladder:

- ❏ If you cannot conveniently reach the work area, move the ladder. If you stretch out to reach something, you could fall or the ladder could tip sideways.
- ❏ Be aware that a tool, such as a drill motor, can hang up and throw you off balance. It can cause a fall.
- ❏ If you work from a ladder, lock yourself to the ladder. Hook one leg through a rung of the ladder. The top of the ladder should be tied off to prevent the ladder from sliding sideways.
- ❏ Do not stand on the top two rungs of the ladder. You have nothing to brace yourself against and the chances of falling are greatly increased.

Metal ladders are durable and strong. However, metal conducts electricity. Never use a metal ladder when working on electrical wiring. Use a wood or fiberglass ladder.

When using **extension ladders**, check to be sure that the rail guides are in good order. Also be sure that the two locking devices that fit over a rung are both in place.

When using **step ladders**, follow these precautions:

- ❑ Be sure the legs are extended fully and the braces are locked straight.
- ❑ The top of the ladder is **not a step**. If you stand on the top, there is nothing to brace your legs against to keep your balance.
- ❑ Do not stand on the paint shelf.
- ❑ Have a solid base. Stepladders usually have small feet. They can sink into the ground easily. Set the ladder on level boards, if there is any doubt. Even a small movement of the ladder can throw you off balance.
- ❑ Don't leave anything lying on the paint shelf or the top of the ladder.
- ❑ Never climb on the diagonal bracing on back of the ladder.

REVIEW

1. Identify the three types of ladder shown:

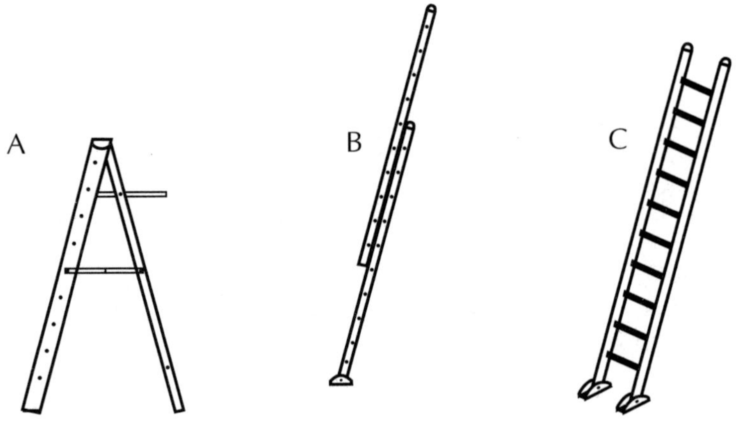

2. Explain the 4 to 1 rule for ladders.

3. Why are painted ladders not used?

4. List three ways to keep a ladder from slipping away from the building.

5. How far should a straight ladder extend above the upper level?

6. What kind of ladder should be used when working on electrical wiring?

7. List at least four rules to be followed when working with step

   ladders.

Check your answers in the back of the book.

10 SCAFFOLDS AND WORK PLATFORMS

Indoor environment technicians—especially sheet metal workers—may work from scaffolds or work platforms.

SCAFFOLDS

Many things can go wrong when a scaffold is being used:

- ❏ Workers fall off.
- ❏ Tools, equipment, or debris fall and hit workers below.
- ❏ Workers receive an electrical shock when the scaffold contacts electrical wires.
- ❏ The scaffold tips over because of the base is unstable or not level.
- ❏ The scaffold tips because it is not tied to the building or otherwise braced.
- ❏ The scaffold collapses because of improper bracing.

Because scaffold accidents are frequent, OSHA has extensive regulations on their erection and use. OSHA lists over 17 different types of scaffolds with regulations for each.

Scaffolds are usually erected by the company that furnishes the scaffolding. Scaffolds must be erected under the supervision of a "trained and competent person" (OSHA). OSHA also specifies that the employer must ensure that scaffolding is safe. However the technician who uses it must know the safety requirements for a scaffold as well as safe work practices on it. **OSHA specifies that employees who work on scaffolds be trained to recognize hazards and to understand how to control the hazards**.

The following section covers general rules for working on all types of scaffolds. The rest of this chapter covers safety regulations that apply to two types of scaffold commonly used—the **tubular frame scaffold** and the **rolling scaffold**. For other types of scaffolds, refer to OSHA regulations.

Working on Scaffolds

- ❑ Never work on a scaffold that you consider unsafe. Any supervisor who asks you to is violating OSHA regulations and is subject to fines.
- ❑ Wear a hard hat.
- ❑ Check to see that no electrical wires can be touched by the scaffold or the workers.
- ❑ Clear the area around the scaffold as much as possible of equipment, materials, and debris. Falling on these objects can make injuries worse.
- ❑ Place warning lines or barricades around the area where there is danger of falling tools or material.
- ❑ If a scaffold work platform is more than 10 feet above the next lower level, workers are required to wear fall protection gear. Working over hazardous conditions such as water or machinery also requires fall protection.
- ❑ Never carry tools or material when climbing a scaffold. Hoist everything by rope.
- ❑ Guard against conditions that can cause tripping and possible falls from the scaffold:

    Keep the scaffold platform clear of tools and debris.

Keep tools stored in a container (such as a bucket) so that they cannot be tripped over or kicked off the platform.

Do not have extension cords lying on the platform.

- ❑ Avoid any projections on the work platform (such as projecting nails).
- ❑ Do not use ladders on scaffold platforms.
- ❑ Do not use makeshift devices (such as boxes or barrels) to gain more height above the work platform.

Tubular Frame Scaffolds

Tubular frame scaffolds (Fig. 1) are usually rented from a specialty company and are erected by that company's employees. These employees are required by OSHA to be trained and experienced in scaffold erection. OSHA also provides that erecting, moving, or dismantling scaffolds must be performed under the supervision of a competent person qualified in scaffold erection.

Fig. 1: Tubular frame scaffold

However, you should not assume that a scaffold you are to work on is properly and safely erected. Your life is at stake. You have the responsibility to inspect a scaffold to see if it is safe to use. Talk to your supervisor if a scaffold seems to be unsafe in any way.

- ❑ The legs of a scaffold must have a firm and level base. The legs must rest on pads or planks that are wide enough to keep the legs from sinking into the ground. Makeshift materials should not be used for pads.

- ❑ The scaffold must be tied to the building or braced to prevent it from tipping over if the vertical height is four times the narrowest width of the base. In other words, a scaffold 4 feet wide must be tied or braced to the building if the top platform is 16 feet above the ground. Then it must be tied at about every 20 vertical feet.
- ❑ Each level of the scaffold frame must have diagonal cross braces in all directions (Fig. 1).
- ❑ A stairway or ladder must be provided as part of the scaffold. You should never have to climb up the outside of a scaffold.
- ❑ Scaffold platforms may be manufactured platforms or scaffold-grade planks:

  Makeshift material, warped planks, or planks with splits in them should not be used.

  Planks must be 2" x 10".

- ❑ The space between a building and the scaffold platform should not be over 14 inches.
- ❑ If scaffold planks are used instead of manufactured platforms, they must be properly set so that they cannot be tipped up by someone stepping on the end of the plank:

  The end of the plank must extend at least 6" but no more than 12" beyond its support (Fig. 2).

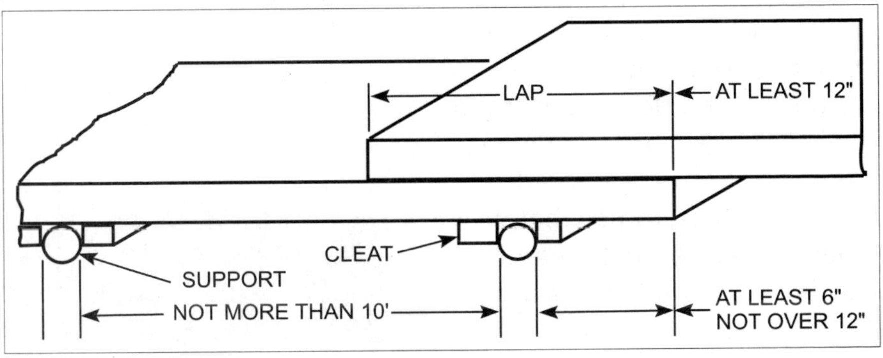

Fig. 2: Planks overlap a support

Planks must have cleats to prevent them from moving on the support (Fig. 2).

Planks can overlap only over a support (Fig. 2). The lap must be at least 12".

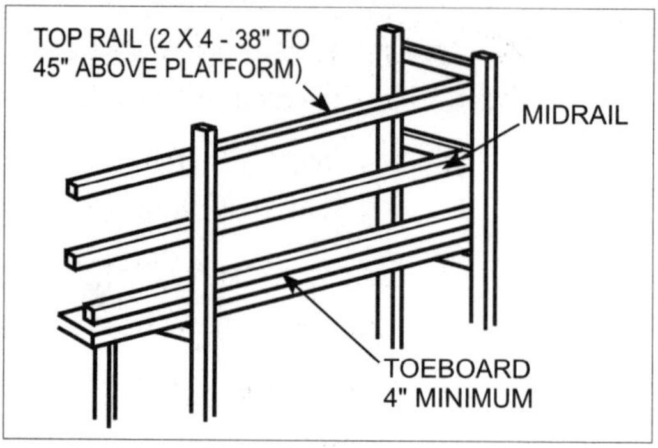

Fig. 3: Guardrail system on a scaffold

- ❑ Every scaffold platform must have toe boards and guardrails on all open sides (Fig. 3). Guardrails must be able to withstand at least 100 pounds side pressure.
- ❑ Never place planks on the top of guardrails to obtain more height.

Rolling Scaffolds

Rolling scaffolds (Fig. 4) are generally used inside a building on a level floor. The indoor environment technician uses them to work on such things as ductwork or suspended HVAC units. OSHA regulations refer to them as **mobile scaffolds**. All the safety regulations for working on scaffolds apply to rolling scaffolds.

Rolling scaffolds are manufactured. They are commonly made of aluminum tubing. The scaffold parts are delivered to the job and the technician assembles them. They assemble quickly and easily. The connections are specially designed for snap-together, positive locking.

COURTESY GENERAL STEEL PRODUCTS CO., INC.

Fig. 4: Rolling scaffold

When erecting and using rolling scaffolds, be sure that all of the following conditions exist:

- ❑ Scaffold platforms must conform to the same regulations as tubular frame scaffolds.
- ❑ Double check each connection to be sure that it is snapped in place.
- ❑ The wheels must have positive wheel and swivel lock.
- ❑ The wheels must be locked at all times except when the scaffold is being moved.
- ❑ A ladder is built into the scaffold. Workers should not climb on the scaffold framework.
- ❑ If the maximum vertical height is over 4 times the narrowest width of the scaffold base, the scaffold must be secured from tipping. On rolling scaffolds, this is usually done by outriggers.
- ❑ On scaffolds lower than the 4 to 1 ratio, outriggers should be used whenever there is a possibility of tipping.
- ❑ Before moving a rolling scaffold, check the following;

  The floor is level and clear of debris, holes, or other obstructions.

  There are no overhead obstructions.

  There are no electric power lines within 10 feet of the path of travel.

  All tools and equipment are removed or secured from falling off the platform.

- ❑ If workers remain on the platform during the move, **all** of the following conditions must also exist:

  The wheels have tires of rubber (or similar material).

  The height of the scaffold is no more than 2 times the narrowest width of the base.

The moving force is exerted low on the base of the scaffold.

WORK PLATFORMS

Work platform (Fig. 5) is a general term for a specially designed platform that raises one or two persons to the desired work level. OSHA regulations refer to them as **aerial lifts**. Follow these safety rules for work platforms:

- ❑ The work platform can **never** be within 10 feet of an electrical line.
- ❑ Workers on work platforms must wear fall protection. OSHA specifies a body belt with a lanyard attached to the work platform or the boom. Note that this is a fall **restraint** system because the safety lanyard is short enough to prevent a worker from going over the rail. The 1998 OSHA regulations permit the use of body belts for **restraint** systems but prohibits them for fall **arrest** systems. A body harness is required for all fall arrest systems.
- ❑ The base for elevating a work platform must be prevented from tipping or moving. This requires such things as outriggers, guy wires, or locking and blocking wheels.

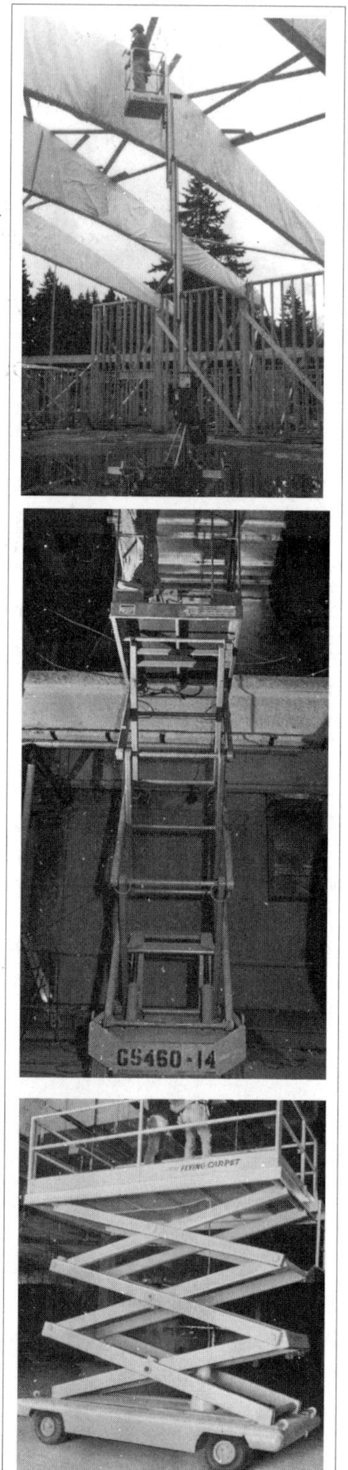

COURTESY GENIE INDUSTRIES

COURTESY UP-RIGHT SCAFFOLDS

Fig. 5: Work platforms

REVIEW

1. When are workers required to wear fall protection gear on a scaffold?

2. What is the required size for scaffold planks?
3. What is the minimum length for a plank to lap over a support?
4. What is the maximum space allowed between a scaffold platform and the building?
5. How far away must electrical wires be from a scaffold?
6. When are outriggers required on rolling scaffolds?

Check your answers in the back of the book.

11 HOISTING

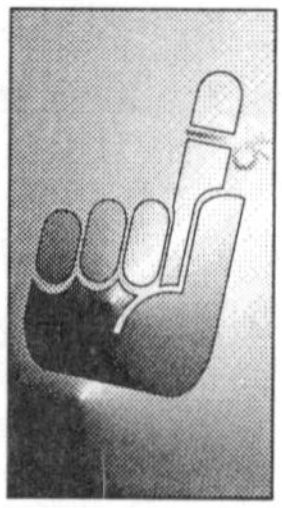

DANGERS IN HOISTING

Hoisting means to raise or lower materials and equipment from one level to another. On new construction, indoor environment technicians—especially sheet metal workers—may be involved when HVAC duct and equipment is delivered to different floors or to the roof. On existing buildings, indoor environment technicians may have to raise or lower materials or equipment to make changes for such things as energy management or indoor air quality.

Unless all safety procedures are followed, hoisting can be very dangerous. These accidents are too common:

- ❏ Workers are injured by falling loads.
- ❏ Structures are damaged when the load swings out of control.
- ❏ A worker receiving the load falls from the roof or floor level.
- ❏ A worker on the ground is crushed between the load and a building.
- ❏ A worker is electrocuted when the crane hits electrical power lines.
- ❏ Equipment or materials being hoisted are damaged in a fall.

Usually there is a crew on the ground that rigs the load so that it can be hoisted. A crew on the roof or upper floor receives the load and settles it in place. Some safety rules apply to anyone working on the hoisting operation:

- ❏ Use only OSHA-approved rigging equipment (wire rope slings, hooks, shackles, etc.). Slings should be prepared by a qualified rigger. Never use makeshift rigging or rigging prepared on the job. Fiber or synthetic ropes should be used only for small loads that are hoisted by hand.
- ❏ Wear gloves when working with wire rope slings. One loose wire can make a bad cut.
- ❏ Wear a hard hat.
- ❏ Do not change rigging that is under load. Stop the hoist and relieve the load if slings or other gear must be adjusted.
- ❏ Personnel are never permitted to ride the load or be carried by equipment designed to carry the load. In very special cases, where there is no other way to reach the work area, OSHA should be consulted for approval to hoist workers by crane.

Ground Operations

The ground crew is responsible for rigging the load safely. This requires an extensive knowledge of such things as the safe use of different types of slings, hooks, shackles, and webbing. It also requires the ability to select the sling arrangement suitable for the load, and to establish the center of gravity so that the load is level. The person in charge of the ground crew should have had a training course in rigging. Improper rigging can easily result in loads falling or slipping:

- ❏ Rope slings, webbing, and rigging attachments should be inspected for damage **at each use**.

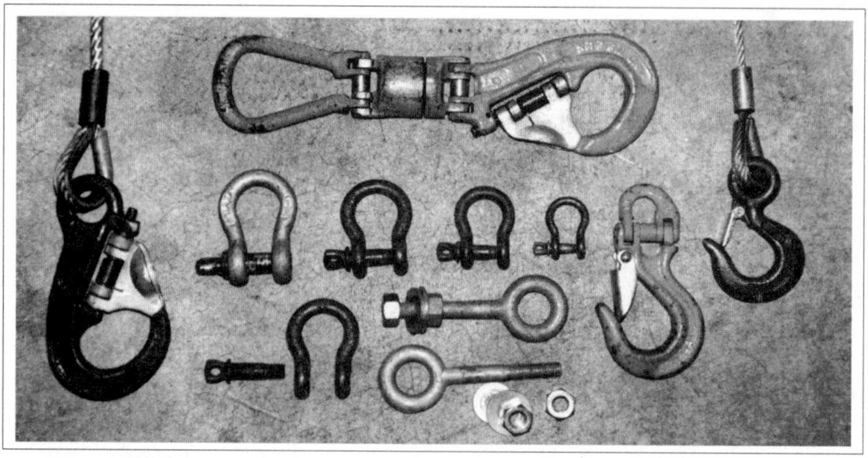

Check rigging equipment to make sure it is in good condition

- ❏ Hooks used for hoisting must have an approved safety latch that prevents the hook from accidentally slipping free.
- ❏ Hoisting hooks should be checked daily to make sure that the hook has not been stretched and that the safety latch operates properly.
- ❏ Keep the hoist area free of debris that can cause tripping. Remove unused rigging equipment from the hoisting area.
- ❏ Put up barricades or warning lines to keep others out of the hoisting area.
- ❏ Never put your fingers between a wire rope and the load.
- ❏ Use cardboard, wood, or other padding between the load and the sling. This protects the sling from damage by the sharp edges of the load, and it protects the edges of the load from damage by the sling.
- ❏ Lift the load slowly to be sure there are no kinks in the wire rope and that all the attachments are located properly. Raise the load slightly to check whether it is hanging level.
- ❏ Never stand under a load or between the load and a wall.

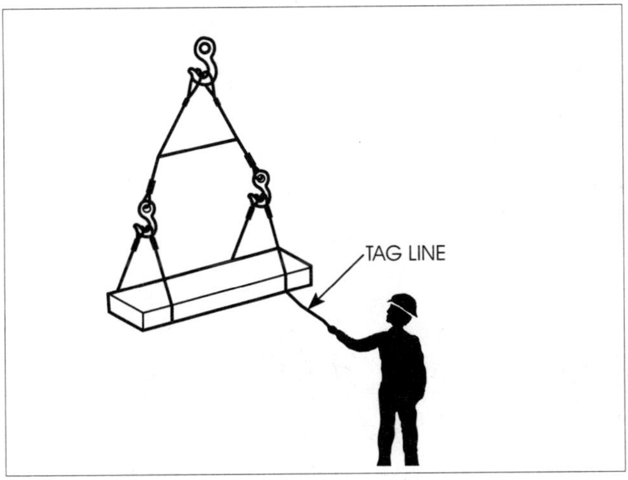

Fig. 1: Use tag lines to guide the load

- ❑ Use tag lines if the load needs to be guided (Fig. 1). Never try to push it by hand.

Top Crew

The top crew receives the load. A member of the top crew signals the hoist operator to guide the load to the desired location. Safe practices must be followed:

- ❑ If the top crew must work near the edge of a roof or floor, fall protection gear, suitable for the situation, must be worn.
- ❑ If slings go under the load, plan the placement of the load so that there is room to remove the slings.
- ❑ The roofing could be damaged when the load is set on the roof, resulting in leaks later on. Protect roofing by setting the load on planks.

Hoisting Signals

Some kind of communication is needed between the crane operator and the ground and top crew. Communication is also needed between the top crew and ground crew when other hoisting equipment is used. Radio communication is commonly used, but sometimes hand signals may be used instead. Figure 2 shows the standard hoisting signals. Everyone working on hoisting should know these.

Fig. 2: Standard hoisting signals

Only one person should give signals at one time. Otherwise, confusion can lead to accidents. Someone on the ground crew gives the signal to lift. Someone on the top crew gives the signals when the load is being raised.

ELECTRICAL HAZARDS

If a crane or other hoisting boom strikes an electric power line, workers can be electrocuted. In some cases, an electrical charge can be induced in hoisting equipment even if there is no direct connect. Many fatal electrical accidents have occurred during hoisting. That is why OSHA has established strict regulations:

- ❑ The hoisting equipment or the load must **always** be **at least 10 feet away from any power line**.
- ❑ If power lines carry over 50,000 volts, the required distance must be **more than** 10 feet. If you work near lines of this very high voltage, be sure to consult OSHA for permissible safe working distances.
- ❑ If the hoisting equipment is being moved, with the **boom lowered** and with **no load** the minimum clearance to a power line is 4 feet. For lines **over 50,000 volts**, the required clearance is 10 feet or more. For these very high voltages, consult OSHA regulations for the proper clearances.
- ❑ If there is the slightest danger of striking power lines during hoisting, have the lines de-energized. Never assume a line is dead unless a representative from the power company certifies that it is. In addition, OSHA requires that the de-energized line be visibly grounded at the point of work.

HELICOPTER LIFTS

Where it is not practical to use hoisting equipment, a helicopter may be used to lift and deliver the load. The helicopter company provides the ground crew or provides training to the workers at the job site.

Some special precautions apply to working with helicopter lifts:

- ❑ A briefing must be conducted between the pilot and the ground personnel before each day's operation. This meeting is to cover such things as safety, procedures, and the method of communication between the pilot and the ground.
- ❑ The helicopter crew should inspect and approve the rigging equipment (cargo hooks, slings, shackles, etc.).
- ❑ Everyone on the site should wear eye protection and a hard hat secured with a chin strap.
- ❑ Avoid loose fitting clothing, which can flap in the rotor downwash and snag on the rigging.
- ❑ Clear the hoist area of debris. If possible, sweep the area for a distance of 100 feet from the lift or the deposit spot. Otherwise the rotor downwash will blow dust and debris in air.
- ❑ Be aware that the downwash can blow any flame or fire out of control.
- ❑ Tag lines to guide the load must be short enough so that they cannot be drawn up into the helicopter rotors.
- ❑ The load will have a grounding line. Before the ground crew touches the load, this line must touch the ground or building in order to discharge any build-up of static electricity.

- ❑ Remember that the load may swing. Do not stand between the load and a wall or between the load and a roof edge.
- ❑ Do not stand under the helicopter except to hook or unhook loads.

REVIEW

1. Why should gloves be worn when handling wire rope?
2. How often should rigging be inspected?
3. What is the purpose of the safety latch on a hook?
4. Why should padding be used between the load and sling?
5. What are two ways of communication between the ground crew and the crane operator?
6. What is the meaning of the signal with the left arm extended horizontally?
7. What does the same signal mean with the hand moving rapidly right and left?
8. On a helicopter lift, why should the tag line be short?
9. What is the purpose of a grounding line?

Check your answers in the back of the book.

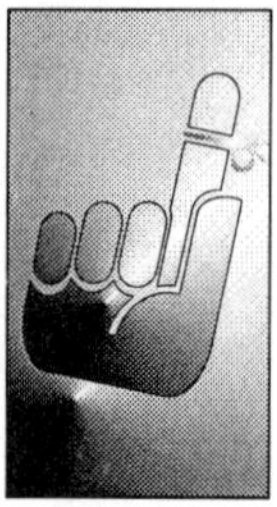

REVIEW ANSWERS

Chapter 1 has no Review.

Chapter 2

Answers to 1 to 3 will vary. They could include any of the following:

1. Asbestos, lead
2. Solvents, gasoline, cements, corrosives, glues, acids
3. Gasoline fumes, gas appliance vents, welding fumes, refrigerant fumes, engine exhaust
4. OSHA
5. Material safety data sheet
6. Hazardous fumes may be present.
   Oxygen may be lacking.
   Gases could explode.
7. Safety monitor
8. Air-purifying respirator
9. Supplied-air respirator
10. Soft ear plugs, ear muffs

Chapter 3

1. Insulators: Dry air, glass, rubber, plastic
   Conductors: Water, silver, aluminum, copper, human body

2. It can cause the circuit to get so hot that it burns through the insulation. Without insulation to keep it contained, electricity could cause shock or fire.

3. Battery, capacitor

4. It is an outlet that turns off the power immediately when a short occurs.

5. Unplug it and tag the plug.

6. Lock out and tag the main power source, the disconnect switch, or the control circuit.

7. The conductor could burn through the insulation and cause shock or fire, and a voltage drop could damage the equipment.

8. It is 911 in most areas. Find out if your area has a different emergency number.

Answers to the rest of the questions may vary. The following are good answers.

9. Turn off the circuit, if possible.
   Use an insulating material to move the wire from the victim.
   Drag the victim away by using a rope, blanket, coat, belt, or other non-conductor.

10. The switch may be wired wrong, so there is still power in the circuit. You may get a shock from a capacitor that is storing electricity.

11. You did not tag the electrical shut-off, and someone turned the power back on.

12. He may have been wet or standing on wet ground so that he made a good conductor. He may have had both hands on a grounded pipe so that the current went through his heart.

Chapter 4

1. Can include any of the following:
   Cracks in heat exchangers
   Poor connections on vents
   Holes in vents
   Blockage of vent
2. LPG (propane or butane)
3. Natural gas or LPG
4. Soapy water or a gas-detecting instrument
5. 5 minutes at least
6. Pull towards you
7. Box end wrench

Chapter 5

1. None
2. Set up a barrier
3. Turn off the power and lock it out
4. Fan, filters, coils
5. Your fingers could be pulled between the V-belt and the sheave and badly injured.
6. Have a cordless phone so you can call for help if needed.

7. 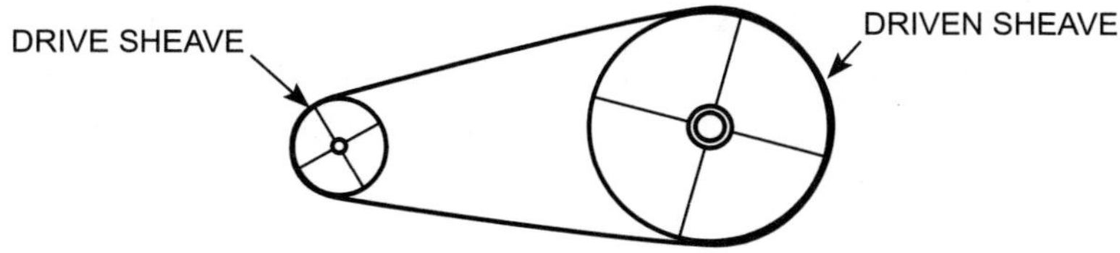

Chapter 6

1. To allow a thermometer or pressure measuring device to be inserted

2. Pressure relief valve

3. Slowly, to avoid shock to the hydronic system

4. An open system is exposed to the atmosphere. A closed system is not.

5. They can spread disease.

6. Crack a valve to relieve pressure.

Chapter 7

Answers will vary.

Chapter 8

1. The location, including the nearest cross streets
   The nature of the accident
   Whether a rescue is needed
   The number of victims
   Approximate condition of the victims and what is being done for them
   Name and phone number of caller

2. Fall restraint gear keeps you from falling.
   Fall arrest gear catches you if you have a fall.

3. Safety harness
   Lanyard
   Lifeline

4. Answers will vary.

Chapter 9

1. A. Stepladder
   B. Extension ladder
   C. Straight ladder

2. The bottom of the ladder should be 1 foot away from the wall for every 4 feet of working length of the ladder.

3. Paint can hide splits and cracks.

4. Tie the foot of the ladder to the building.
   Tie the top of the ladder to the building.
   Nail a cleat at the foot of the ladder.

5. 3 feet

6. Wood or fiberglass

7. Answers will vary and can include any of the following:
   - Be sure the legs are extended fully and the braces are locked straight.
   - Do not stand on the top of the ladder.
   - Do not stand on the shelf.
   - Have a solid base.
   - Don't leave anything lying on the shelf or the top of the ladder.

Chapter 10

1. When the scaffold is more than 10 feet above the next lower level

2. 2" x 10"

3. 6"

4. 14"
5. Far enough so that they cannot be touched by the scaffold or by workers on it.
6. When the vertical height is over 4 times the narrowest width of the scaffold base, or if there is danger of tipping.

Chapter 11

1. Because loose wires can cut your hands
2. Every time it is used
3. To prevent the hook from slipping free
4. To protect the load and to protect the sling
5. Radio or hand signals
6. Stop
7. Emergency stop
8. To keep it clear of the rotors
9. To discharge any static electricity in the load

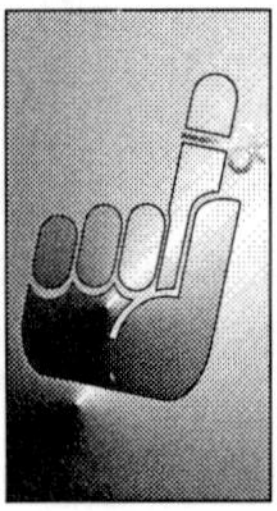

Index

LAMA Books also publishes

Teach! Plain Talk About Teaching

124 pages, illustrated
7 x 9 softcover - $20.00
ISBN 0-88069-014-3

Teach is an excellent source for anyone teaching.

THIS WILL HELP YOU:

- Teach practical subjects to adults
- Lead dynamic class discussions
- Manage classroom distractions
- Engross students in learning
- Use media aids effectively
- Write tests and quizzes

The Indoor Environment Technician's Library

Basics of Electricity • Airflow in Ducts • Math for the Technician
Variable Air Volume Systems • Indoor Air Qualtiy
Safety for the Indoor Environment Technician

100 approx. pages, 6 x 9 softcover - $18.50
Subscribe to the IET series and get the books for $15.00

More than just reference books, the series marks a new way to address the training problems of the industry. Each book focuses on the skills for the technician in heating, ventilating, and air conditioning. May be used in supervised training programs, self-training situations, or just for reference. Books are a handy size, easy reading, with a review at the end of each chapter.

Two different references that cover all the occupational programs in California and the western states.

Occupational Programs in California Community Colleges

176 pages 8.5 x 11 softcover
$30.00

Occupational Programs in the Western States

210 pages 8.5 x 11 softcover
$10.00

Check us out on the Web!
WWW.LAMABOOKS.COM

ORDER FORM

Name ____________________

Title ____________________

School/Business ____________________

Street Address ____________________

City/State/Zip ____________________

Telephone ____________ Fax ____________ email ____________

IET *Indoor Environment Technician's Library* (Choose the one you would like to start out with or get any one you wish.) I understand that future titles will be sent periodically on approval and I may return them if they do not fit my purposes. ☐ Please enroll me as a subscriber to the IET Library @ a 20% discount ☐ Please send me more information on the IET Library		#	Total
IET titles available: *Basics of Electricity*	$18.50 non member/$15.00 member's		
Airflow in Ducts			
Math for the Technician			
Variable Air Volume Systems			
Indoor Air Quality			
Safety for the Indoor Environment Technician			
2001 - 2002 Occupational Programs in California Community Colleges	$30.00		
Occupational Programs in the Western States	$10.00		
TEACH! Plain Talk About Teaching	$20.00		
The Reading Program, Books A to G	$8.00 each		
Tests and Checkpoints	$15.00		
Tapes for Book A	$35.00		
	Sub total		
	California residents add appropriate sales tax		
	Add 7% shipping & handling		
	TOTAL		

PAYMENT:

☐ Check
☐ Purchase order
☐ Bill me

LAMA Books
Leo A. Meyer Associates Inc.

20956 Corsair Blvd • Hayward CA 94545-1002 • 510-785-1091 • 888-452-6244 • 510-785-1099 Fax • lama@lamabooks.c

www.lamabooks.com